Hydrothermal Technology in Biomass Utilization & Conversion

Hydrothermal Technology in Biomass Utilization & Conversion

Special Issue Editors

David Chiaramonti

Andrea Kruse

Marco Klemm

MDPI • Basel • Beijing • Wuhan • Barcelona • Belgrade • Manchester • Tokyo • Cluj • Tianjin

Special Issue Editors
David Chiaramonti
Renewable Energy Consortium for Research and Demonstration
Italy

Andrea Kruse
University of Hohenheim
Germany

Marco Klemm
DBFZ Deutsches Biomasseforschungszentrum gemeinnützige GmbH
Germany

Editorial Office
MDPI
St. Alban-Anlage 66
4052 Basel, Switzerland

This is a reprint of articles from the Special Issue published online in the open access journal *Energies* (ISSN 1996-1073) (available at: https://www.mdpi.com/journal/energies/special_issues/Hydrothermal_Technology_Biomass_Utilization_Conversion).

For citation purposes, cite each article independently as indicated on the article page online and as indicated below:

LastName, A.A.; LastName, B.B.; LastName, C.C. Article Title. *Journal Name* **Year**, *Article Number*, Page Range.

ISBN 978-3-03928-676-8 (Pbk)
ISBN 978-3-03928-677-5 (PDF)

Contents

About the Special Issue Editors

David Chiaramonti Polytechnic of Turin, Energy Department "Galileo Ferraris", Corso Duca degli Abruzzi 24, I-10129 Turin, and Re-cord – Renewable Energy Consortium for Research and Demonstration, Italy. Main Interests: hydrothermal carbonization and liquefaction; pyrolysis; biocrude upgrading; biochar and biochar-derived products characterization and use; bio and thermochemical process integration; biofuels and biopoducts.

Andrea Kruse Institute of Agricultural Engineering, Conversion Technologies of Biobased Resources, Universität Hohenheim / University of Hohenheim, Stuttgart, Germany. Interests: hydrothermal carbonization; carbon materials; platform-chemicals from biomass; nutrient recovery; hydrothermal conversion; hydrothermal liquefaction; hydrothermal gasification; hydrothermal pretreatment.

Marco Klemm DBFZ Deutsches Biomasseforschungszentrum gemeinnützige GmbH, Leipzig, Germany. Interests: hydrothermal processes for solid and liquid products for different applications (e.g., solid fuel, carbon materials, liquid fuels, chemicals); balance, technical assessment, and optimization of the hydrothermal process; implementation of hydrothermal processes in provision chains; assessment of feedstock concerning the application in hydrothermal processes.

Preface to "Hydrothermal Technology in Biomass Utilization & Conversion"

In recent years hydrothermal processing (HTP), in particular, liquefaction (HTL) and carbonization (HTC), came into the spotlight for scientists, process developers, institutions and investors, even if HTP itself originated many years ago. The growing interest in biomass conversion to energy and chemicals (linked to climate targets), combined with the abundance of wet streams that could be efficiently converted into fuels or products and the diffusion of the circular economy concept, which aims at valorizing all streams towards zero-waste configurations, ranked this process in the priority list for research and industrial deployment. Some residual feedstocks or waste streams are, in fact, perfect materials for HTP, such as urban wastewater sludges, among others, of which the management today represents a major issue. However, much in this field remains to be explored and understood, such as the optimal combination of feedstocks and process conditions, or how to drive the process of specific types of products, or the scale-up of the technology to industrial and commercial level, etc. This Special Issue collects some of the latest results from research in a single book: it offers a sound contribution to the literature in the field and the advancement of knowledge in hydrothermal processing. The original research works here included range from the valorization of wet streams, such as from microalgae or lignin-rich streams from lignocellulosic ethanol, to nutrient recovery, extractions, characterizations of carbonized products, technology diffusion and LCA analysis, thus covering most of the areas relevant for HTP. Indications for further research are given, supporting future investigations in the field.

David Chiaramonti, Andrea Kruse, Marco Klemm
Special Issue Editors

Article

Lignocellulosic Ethanol Biorefinery: Valorization of Lignin-Rich Stream through Hydrothermal Liquefaction

Edoardo Miliotti [1], Stefano Dell'Orco [1,2], Giulia Lotti [1], Andrea Maria Rizzo [1], Luca Rosi [1,3] and David Chiaramonti [1,2,*]

[1] RE-CORD, Viale Kennedy 182, Scarperia e San Piero, 50038 Florence, Italy; edoardo.miliotti@re-cord.org (E.M.); stefano.dellorco@unifi.it (S.D.); giulia.lotti@re-cord.org (G.L.); andreamaria.rizzo@re-cord.org (A.M.R.); luca.rosi@unifi.it (L.R.)

[2] Department of Industrial Engineering, University of Florence, Viale Morgagni 40, 50135 Florence, Italy

[3] Chemistry Department "Ugo Schiff", University of Florence, Via della Lastruccia, Sesto Fiorentino, 50019 Florence, Italy

[*] Correspondence: david.chiaramonti@re-cord.org; Tel.: +39-055-2758690

Received: 18 January 2019; Accepted: 15 February 2019; Published: 22 February 2019

Abstract: Hydrothermal liquefaction of lignin-rich stream from lignocellulosic ethanol production at an industrial scale was carried out in a custom-made batch test bench. Light and heavy fractions of the HTL biocrude were collected following an ad-hoc developed two-steps solvent extraction method. A full factorial design of experiment was performed, investigating the influence of temperature, time and biomass-to-water mass ratio (B/W) on product yields, biocrude elemental composition, molecular weight and carbon balance. Total biocrude yields ranged from 39.8% to 65.7% *w*/*w*. The Temperature was the main influencing parameter as regards the distribution between the light and heavy fractions of the produced biocrude: the highest amount of heavy biocrude was recovered at 300 °C, while at 350 and 370 °C the yield of the light fraction increased, reaching 41.7% *w*/*w* at 370 °C. Instead, the B/W ratio did not have a significant effect on light and heavy biocrude yields. Feedstock carbon content was mainly recovered in the biocrude (up to 77.6% *w*/*w*). The distribution between the light and heavy fractions followed the same trend as the yields. The typical aromatic structure of the lignin-rich stream was also observed in the biocrudes, indicating that mainly hydrolysis depolymerization occurred. The weight-average molecular weight of the total biocrude was strictly related to the process temperature, decreasing from 1146 at 300 °C to 565 g mol^{-1} at 370 °C.

Keywords: lignin; biorefinery; hydrothermal liquefaction; biocrude; depolymerization

1. Introduction

The EU-Renewable Energy Directive (RED) defines Advanced Biofuel only on the base of the feedstock (as reported in Annex IX Part A of RED [1]). The use of residual/dedicated lignocellulosic biomass is currently promoted for sustainable biofuel production, a sector that is largely dominated by lipids in Europe. Vegetable and used oils, representing by far the largest share of biofuels in EU [2], even when converted into high-quality hydrocarbons through hydrotreatment and hydroisomerization processes, are criticized for the potential food versus fuel conflicts and the use of high-ILUC (indirect land use change) feedstock, such as imported palm oil. The major Research and Development efforts in the EU focus on developing new industrial-scale technologies able to produce sustainable biofuels/bioenergy from lignocellulosic material. The lignocellulosic ethanol route is among these pathways. It has achieved full industrial scale worldwide as a consequence of the wide diffusion of this process, an increasing amount of a very wet lignin-rich stream (LRS) is made available as co-product at the production site in considerable quantities, constant physical and chemical characteristics, and affordable costs [3]. The current use of the LRS in industrial complexes is still limited to combustion for heat and power generation. However, being lignin the most abundant renewable source of aromatics in nature, its valorization is a very attractive opportunity for green chemistry in a circular economy perspective. Therefore, several research works addressed the economic valorization of lignin-rich streams from lignocellulosic ethanol production, either as chemical or as fuel, highlighting the challenges and importance of co-product valorization to achieve commercial competitiveness of these processes [4,5]. The economic relevance of lignin co-products valorization clearly emerged since the initial modelling studies of the process [6] and it was later confirmed by experimental data from a pilot, demo and first-of-a-kind plants.

Among the different processes and technologies that deal with lignin depolymerization [7], hydrothermal liquefaction (HTL) is a noteworthy thermochemical process which can convert lignocellulosic biomass mostly into a liquid fraction by using solely hot compressed water, or mixtures of water, co-solvents and chemicals [8,9]. HTL is a wet process, which does not require feedstock drying, as it is instead necessary for other thermochemical processes like gasification and pyrolysis. As such, HTL is an attractive approach for the conversion of wet biomass into a liquid product. Therefore, the high water-content, rather a constant composition, and the continuous availability at the industrial site of the lignin-rich co-product makes it a promising candidate for processing under hydrothermal liquefaction conditions into a biocrude. This would significantly improve the overall biorefinery carbon efficiency and economic performances, opening new business opportunities.

Several authors carried out fundamental investigations on HTL of lignin using model compounds, as Vanillin, Monobenzone and 2-2′-biphenol [10]. They showed that ether bonds are more reactive under hydrothermal conditions than C-C bonds. Thus, the liquid yield reduces from monobenzone (almost complete conversion) to vanillin to 2-2′biphenol (minimum conversion). Both fragmentation and condensation reactions occur on phenolic compounds in a hydrothermal environment, probably in competition, depending on the specific conditions. During hydrothermal liquefaction of lignin, α- and β-aryl ether hydrolysis, C—C bonds cleavage, alkylation, deoxygenation and repolymerization reactions take place simultaneously, whereas typically the aromatic structure is not affected by hydrothermal reactions. High molecular weight compounds from lignin HTL come from the partial depolymerization of the initial lignin from selective ether bonds splits but also from alkylation of the aromatic structures.

HTL conversion of lignin stream is often carried out at 350–400 °C, 22 MPa and 10 min residence time [11]. The process generates an energy-dense biocrude as the main fraction, along with gaseous products, solids, and an aqueous-phase byproduct. The biocrude yields can reach typically around 40%–50% w/w, with catechol, phenols, and methoxyphenols as main constituents. Similar results are obtained in the hydrothermal treatment of Kraft pine and organosolv lignin [12]. Most of the known HTL studies addressed lignin from pulp and paper or high-purity model compounds [10–15], both of them structurally different from lignin-rich stream originated from lignocellulosic ethanol biorefineries, which is still an unexploited material. To the best of authors' knowledge, the study publicly available, which shares the closest similarities with the feedstock reported in the present work is due to Jensen et al. [16], who studied the influences of pre-treatment on the product composition for the case of alkaline HTL of a lab-scale, lignin-rich enzymatic hydrolysis residue.

The present work investigates the conversion of a lignin-rich stream from industrial-scale lignocellulosic ethanol into a biocrude suitable for further processing and upgrading into fuels and chemicals. The feedstock considered here is thus the actual stream from the industrial process. A special focus was given to the development and implementation of an extraction method in combination with the batch HTL micro-reactors system used in this research.

2. Materials and Methods

2.1. Lignin-Rich Stream

The lignin-rich stream (LRS) was obtained after ethanol distillation and mechanical separation of water from a demo lignocellulosic ethanol plant fed with poplar. The feedstock was dried in an oven for 48 h at 75 °C, knife-milled and then sieved to 0.25 mm; the LRS arrived as a moist agglomerated powder. After drying these agglomerates were size-reduced by milling. The characterization of this feedstock is given in the results and discussion section.

2.2. Experimental Equipment and Procedure

Batch hydrothermal liquefaction experiments were carried out in a custom-made micro-reactor test bench (MRTB), described in a previous publication by the authors [17]. The reactor consists of an AISI 316 $\frac{3}{4}$ (outer diameter) tube with a length of 300 mm (~43 mL of internal volume). In order to prepare batch experiments, dried feedstock was dispersed in ultrapure water (0.055 μS cm^{-1}) to attain the desired biomass-to-water mass ratio. The mass of slurry loaded into the reactor was 33 g for each test. Prior to each experiment, a leakage test was performed with argon pressurized at 8 MPa. Then, three purging cycles with nitrogen (0.5 MPa) were carried out in order to remove air and ensure an inert atmosphere in the reactor. An initial pressure of 3 MPa was set using argon, then the reactor was immersed into a fluidized sand bath. Counting of residence time started when the inner reactor temperature reached 2 °C below the set reaction temperature: as the design residence time was completed, the reactor was rapidly cooled by immersion in a water bath. After nearly 20 min, the pressure was gradually released, the reactor opened and disconnected from the test bench.

A full factorial experimental plan with three factors and two levels was adopted, and the influence of temperature, time, the biomass-to-water mass ratio (B/W) and their interactions on the biocrude yield was assessed by means of an analysis of variance (ANOVA) on the experimental results. Each experiment was replicated between two to three times. In Table 1, the factors and the related low and high levels are reported. Besides the experiments planned, higher temperature (370 °C) and longer residence time (15–20 min) were also investigated in order to find the maximum yield of light biocrude.

Table 1. Operating parameters of the design of the experiments (DOE).

Factor	Low Level	High Level
Temperature (°C)	300	350
Time (min)	5	10
B/W (-)% w/w d.b. [1]	10	20

[1] d.b.: dry basis.

In the present study, a light and a heavy fraction of the biocrude, named biocrude 1 (BC1) and biocrude 2 (BC2), respectively, were recovered with a two-steps solvent extraction method. The selection of the solvent for the recovery of the light fraction (diethyl ether, in short DEE) was based on a comparison with dichloromethane (GC-MS analysis). The solvent for the extraction of the heavy fraction (acetone, in short DMK) was based on the literature (see Appendix A for detailed results and references). Two different collection procedures were first developed and then evaluated, named Procedure 1 and Procedure 2, whose block diagrams are shown in Figure 1. In regards Procedure 1, once the reactor is disconnected from the test bench, it is rinsed with DEE and its content is vacuum-filtered over a Whatman glass microfiber filter (1 μm). Water and water-soluble organics (WSO) are then recovered by gravity separation, while biocrude 1 is obtained after rotary evaporation of DEE at reduced pressure. The reactor and the solids are then rinsed with DMK; then, the DMK and the DMK-solubles are subjected to rotary evaporation at reduced pressure for the collection of biocrude 2, while the solid residue (SR) is oven-dried at 105 °C overnight. Procedure 2 differs only in the first step, where water and WSO are collected prior to solvent extraction. The products obtained from the experiments defined in the DOE were collected according to Procedure 1.

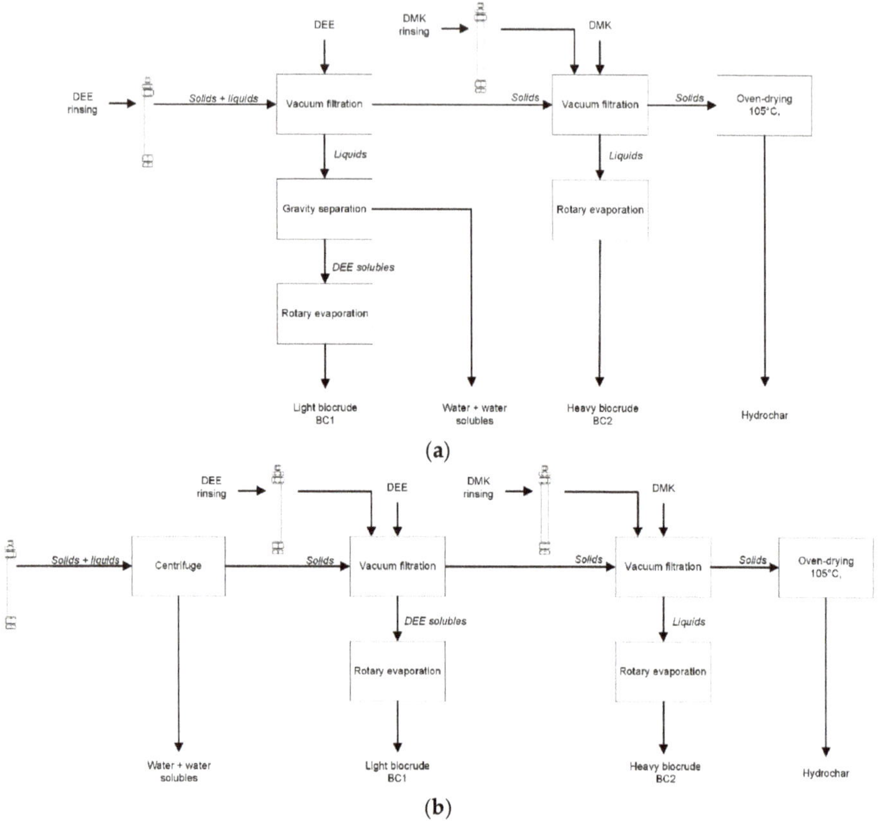

Figure 1. Scheme of Procedure 1 (**a**) and 2 (**b**) for products collection.

The mass yield and the carbon yield of the HTL products were evaluated according to Equations (1) and (2) below

$$\text{Mass yield} = \frac{\text{mass of product}}{\text{dry mass of LRS}} \times 100 \quad (1)$$

$$\text{Carbon yield} = \frac{\text{carbon mass in the product}}{\text{carbon mass in the LRS (d.b.)}} \times 100 \quad (2)$$

The gas yield was estimated assuming the produced gas fraction as composed entirely by CO_2, and ideal gas behavior, while the unrecovered products and the WSO fraction, which were not detected in HPLC, were determined by difference. The assumption of considering the gas phase made up entirely of carbon dioxide is legitimated, under a reasonable degree of approximation, by the fact that decarboxylation is one of the main reaction involved in hydrothermal liquefaction, leading to the formation of a CO_2-rich gas [18,19]. The products obtained from a typical experiment are shown in Figure 2. BC1 is a viscous brown liquid, while BC2 was recovered as a powder or as a very viscous black liquid, as similarly experienced by the research group of Xu (Ahmad et al. [20], Cheng et al. [21]) in their experiments on hydrothermal depolymerization of lignin.

Figure 2. Products collected from a typical hydrothermal liquefaction (HTL) experiment in the micro-batch reactors: (**a**) light biocrude or BC1; (**b**) heavy biocrude or BC2; (**c**) aqueous phase with water-soluble organics (WSO); (**d**) solid residue.

2.3. Analytical Methods and Chemicals

Prior to feedstock characterization, the LRS was dried at 75 °C for 48 h and milled in a knife mill (RETSCH SM 300) equipped with a 0.25 mm sieve. The drying process was carried out at low temperature in order to minimize the devolatilization of the organic matrix. Moisture, ash content and volatile matter were determined in a Leco TGA 701 instrument according to UNI EN 13040, UNI EN 14775 and UNI EN 15148, respectively. Fixed carbon was calculated by difference. The content of carbon, hydrogen, nitrogen (CHN) was determined through a Leco TruSpec according to UNI EN 15104, while the sulphur content of the feedstock was analyzed by means of a TruSpec S Add-On Module, according to ASTM D4239. The oxygen content was evaluated by difference, considering C, H, N, S and ash content. For the biocrude samples, the sulphur content was neglected in the evaluation of oxygen. Higher heating value (HHV) was measured according to UNI EN 14918 by means of a Leco AC500 isoperibol calorimeter. The HHV of the biocrudes was also estimated with the Channiwala and Parikh equation [22], due to the small available amount of samples. The validity of the latter correlation was assessed by a comparison with the measurement of the HHV of two light and heavy biocrude samples. Details are reported in Appendix D. The determination of the pH of the LRS was performed according to DIN ISO 10390.

The lignin content of the LRS was evaluated by a combination of three NREL procedures:

- The LRS was subjected to Soxhlet extraction with water and then ethanol in order to obtain the water-soluble and ethanol-soluble extractives (NREL procedure TP-510-42619 [23])
- The remaining solid residue was subjected to acidic hydrolysis for the evaluation of the acid soluble, acid insoluble lignin and structural sugars (cellulose and hemicellulose) by UV-VIS spectrophotometer and HPLC (NREL procedure TP-510-42618 [24])
- The ash content of the acid insoluble lignin was measured in order to determine the correct value of the latter (NREL procedure TP-510-42622 [25])

Infrared analyses were carried out with a Fourier transform infrared spectrophotometer (FT-IR, Affinity-1, Shimadzu), equipped with a Specac's Golden Gate ATR.

The evaluation of the apparent molar mass (polystyrene equivalent) of the BC1 and BC2 was determined by gel permeation chromatography (GPC). The samples were firstly dissolved in tetrahydrofuran (THF), left overnight and then passed through a 0.45 µm syringe filter. Afterwards, 100 µL of sample was injected in an HPLC apparatus (Shimadzu LC 20 AT Prominence) connected to a refractive index detector (RID) and equipped with two in-series columns (Agilent, PL gel 5 µm 100 Å 300 × 7.5 mm) and a guard column (Agilent, PL gel 5 µm 50 × 7.5 mm). The analyses were performed at 40 °C with 1 mL min^{-1} of THF as eluent. Linear polystyrene standards (Agilent) with a molecular weight ranging from 370 to 9960 g mol^{-1} were used for calibration.

Qualitative and quantitative analysis of the organic compounds in the light biocrude samples were performed by GC-MS: 2 µL of BC1: isopropanol solution (0.1 g:10 mL) was injected in a GC 2010 with a GCMS-QP2010 mass spectrometer (Shimadzu) equipped with a ZB-5 MS Phenomenex column (30 m length, internal diameter 0.25 mm, film diameter 0.25 µm). The temperature was held at 40 °C for 10 min and then increased to 200 °C (heating rate 8 °C min^{-1}, holding time 10 min) and 280 °C (heating rate 10 °C min^{-1}, holding time 30 min). The qualitative analysis was performed comparing the mass spectra to the NIST 17 library after a previous 4-point calibration with the main compounds observed in the prior qualitative screening, using o-terphenyl as an internal standard.

The concentration of the WSO in the aqueous phase was evaluated by HPLC (LC-20 AT Prominence Shimadzu) equipped with a refractive index detector, a Hi-Plex H column 300 × 7.7 mm (Agilent) and a guard column PL Hi-Plex H 50 × 7.7 mm (Agilent), operating at 40 °C with a flow of 0.6 mL min^{-1} with 0.005 M sulfuric acid as mobile phase. Twenty five microliter of each aqueous sample was injected after a 0.2 µm syringe filtration. The quantitative analysis was accomplished after a 5-point calibration following the NREL 42623 guidelines [26]. In addition, a Karl Fischer titration (848 Titrino Plus, Metrohm) was performed following ASTM E203-08 to determine the WSO yields.

The total organic carbon (TOC) of the aqueous phase was determined by a Merck TOC test kit and a Shimadzu UV-1800 spectrophotometer (605 nm). Samples were heated in a Merck TR320 thermoreactor for 2 h at 120 °C and then allowed to cool for 1 h in a test tube rack at room temperature. As DEE is slightly soluble in water, the TOC measurement of the aqueous samples collected with Procedure 1 was corrected with the method reported in Appendix C.

All solvents and reagents required for this work were purchased from Carlo Erba and Sigma Aldrich: they were used as received without any further purification. All chemicals were ACS reagent grade. Water for HPLC and THF for GPC were HPLC grade. Ultrapure water (0.055 µS cm^{-1}) for HTL experiments was produced with a TKA Microlab ultrapure water system. Analytical standards for GC and HPLC were $\geq$ 98% purity. Chemical standards for HHV and CHNS calibrations were purchased from Leco. All gases were purchased from Rivoira. Argon, air, nitrogen and oxygen were supplied with a 99.999% purity, whilst helium was at 99.9995%.

The statistical analysis for the determination of significant operating parameter was carried out with the software Minitab (Minitab Inc.), by considering a significance level of 5%.

3. Results and Discussion

3.1. Feedstock Characterization

Table 2 reports the properties of the feedstock. As it was obtained after mechanical dewatering, the LRS still has a high moisture content, nearly 70% w/w (w.b.), while its ash content is relatively low, as the fermentation feedstock was a hardwood (poplar) and not a herbaceous biomass, for instance. The lignin content is nearly 54% w/w (d.b.).

The detailed results from the analysis of the lignin content are reported in Table 3: 97.4% of the lignin contained in the feedstock was acid insoluble. After the Soxhlet extraction of the extractives, the residual lignin, cellulose and hemicellulose were approximately ash-free; the low amount of ashes from the LRS were concentrated in the extractives due to leaching during the extraction process. A very low amount of cellulose and hemicellulose (structural sugars) was detected, indicating that these compounds were effectively converted into ethanol during poplar fermentation. The mass balance was very well closed (94.62%).

Table 2. Characterization of the lignin-rich stream (w.b.: wet basis; d.b.: dry basis).

Parameter	Value
Moisture (% w/w) w.b	69.7
Ash (% w/w) d.b	2.6
Volatile matter (% w/w) d.b.	71.0
Fixed Carbon (% w/w) d.b	26.4
Higher Heating Value (MJ kg^{-1})	22.9
C (% w/w) d.b	54.2
H (% w/w) d.b	5.9
N (% w/w) d.b	1.0
S (% w/w) d.b	0.2
O (% w/w) d.b	36.1
Lignin content (% w/w) d.b	53.9
pH (-)	4.4

Table 3. Results from the lignin content evaluation.

Parameter	Value (% w/w) (d.b.)
Water extractives [1]	13.62
Ethanol extractives [1]	24.95
Total extractives	38.57
Acid insoluble lignin	52.51
Acid soluble lignin	1.40
Lignin ashes	b.q.l.
Total lignin	53.91
Structural sugars	2.14
Total	94.62

[1] Ash contribution included; b.q.l.: below the quantification limit.

3.2. Comparison of Extraction Procedures

Given the lab-scale size of the experimental apparatus, the recovery of the HTL products is a challenging task, as some can be retained in the reactor wall after the experiments. In order to collect the largest amount of biocrude from these small reactors, a solvent extraction procedure was developed; it is technically not possible to separate the biocrude and the aqueous phase gravimetrically. This would instead be the preferred solution in case of large scale continuous processes and the same approach should be considered also in lab-scale experiments, as reported also by Castello, Pedersen and Rosendahl [9]

Figure 3 reports the effects of the collection procedure on the composition of the light biocrude fraction (biocrude 1) and on the aqueous phase obtained from an experiment performed at 350 °C, 10 min, 10%. It is clearly visible that by using Procedure 1, the light biocrude has a higher amount of organics and, in particular, catechol, creosol, acetic acid, benzoic acid and 4-ethylguaiacol are under the detection limit in the case of Procedure 2. Accordingly, in the aqueous phase, the situation is reversed: a greater concentration of organics is obtained in the sample collected through the Procedure 2; this is true for all the calibrated compounds, except for lactic acid, glycerol and glycolic acid, whose concentrations are comparable. In addition, Table 4 shows the difference in products yield between the two collection procedures: a higher amount of BC1 and a lower amount of WSO are recovered by means of Procedure 1. This behavior is explained by the fact that in Procedure 1 water is not removed prior to DEE extraction of BC1 and therefore water-soluble organics are in part recovered in the light biocrude. From now on, the results showed in this study were based on this latter collection procedure, which was adopted because it allowed for a larger recovery of organics in the biocrude. However, it should be kept in mind that Procedure 2 would be more suitable for a direct comparison with a scaled-up/continuous process, where the biocrude would be gravimetrically separated from the water.

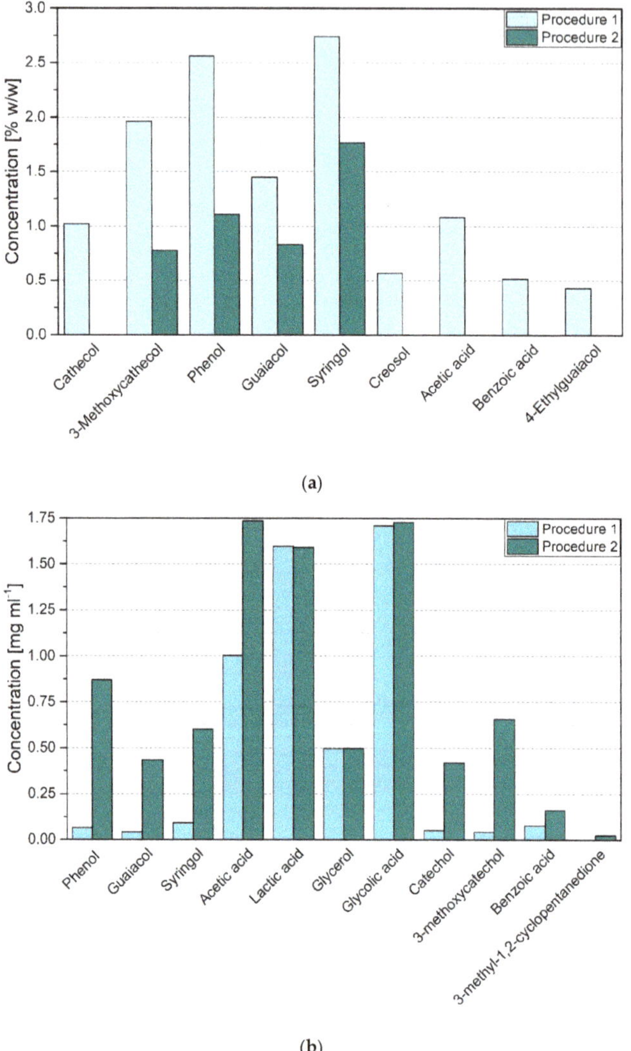

Figure 3. Effect of the collection procedure on biocrude 1 composition by GC-MS analysis (**a**) and on aqueous phase composition by HPLC analysis (**b**) The experiment was performed at 350 °C, 10 min, 10%.

Table 4. Effect of the collection procedure on measured yields of products—experiment performed at 350 °C, 10 min, 10%. Absolute standard deviation is given in brackets.

Product	**Yield (% *w*/*w*) d.b.**	
	Procedure 1	**Procedure 2**
Biocrude 1	29.31 (0.01)	23.1 (1.7)
Biocrude 2	22.5 (6.4)	17.1 (0.7)
Solid residue	11.8 (0.2)	12.7 (0.9)
WSO	12.2 (n.d.)	20.5 (n.d.)
Gas	5.5 (0.9)	5.5 (1.4)

n.d.: not determined.

3.3. *Yields and Influence of Operating Parameters*

Figure 4 shows the yield of the HTL products, which were obtained at the operating conditions selected according to the experimental plan. The unidentified WSO were evaluated by difference and take into account also the losses due to the collection procedure.

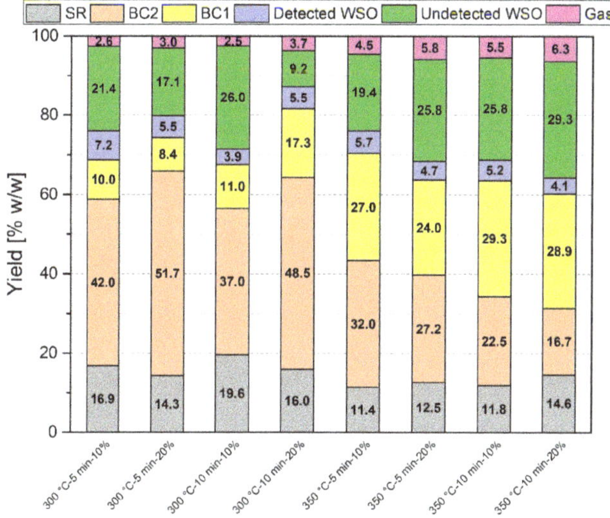

Figure 4. Dry-basis mass yields under different reaction conditions.

A high yield of total biocrude was obtained, ranging from 44.1% to 65.7% w/w, with the amount of light and heavy fraction changing with reaction conditions. In general, by increasing the reaction temperature, an increase in the yield of BC1 and a decrease in that of BC2 are observed, while the solid residue is approximately constant throughout all operating conditions, being char yields between 11.4% and 19.6% w/w. The maximum total biocrude yield was achieved at 300 °C, 10 min, 20% but nearly 74% of it was composed by BC2. At 350 °C, 10 min, 10%, the total biocrude yield was 51.8% w/w and the maximum BC1 yield was obtained (29.3% w/w). The yields of the detected WSO and of the gas products were lower and the latter experienced an increase at 350 °C, as a higher temperature is known to enhance gasification reactions [27]. It is known from the literature [18] that the hydrothermal liquefaction of lignin is more likely to produce a rather high amount of solid product and therefore the use of alkali catalysts, such as KOH, K_2CO_3 or NaOH [27,28], and capping agents as phenol or boric acid [15,28–30] have been suggested to limit the char formation hampering polymerization, as well as different reaction medium than just water, as ethanol, methanol or water-mixture thereof [21,31,32]. For instance, Arturi et al. [30] investigated the effect of phenol in the HTL of Kraft pine lignin with K_2CO_3 and, in the temperature range of 280–350 °C, at a concentration of 3.2%–3.6% w/w of phenol obtained comparable solid yields to the present study, where no additives were adopted and with the use of a similar solvent extraction procedure.

A statistical analysis was also performed in order to assess the influence of process parameters (temperature, time, B/W) and their interaction on BC1, BC2 and total biocrude yield. The significance level for this model was chosen to be 0.05 (95% confidence level). A Pareto plot [33] is reported in Figures 5–7 to visually highlight the absolute values of the main factors and the effect of interaction between the three parameters. The reference line in the chart indicates the limit between significance. The Pareto plot is useful to discriminate which process parameters can be neglected and which ones have an importance in the hydrothermal conversion process. However, to have a deeper understating of the positive and negative effects, the main effects plot and the interactions plot are reported in

Appendix B (Figures A1–A3), showing how positively or negatively each parameter or combination thereof affects the biocrude yield.

It can be noticed from Figure 5 that temperature (A) had the greatest effect on BC1 yield, but also the residence time (B) is statistically significant at 95% confidence level, being both above the mentioned reference line. On the contrary, the ratio B/W (C), as well as the combination of factors, can be considered as not significant for the yield of BC1. This validates the fact that the temperature and partially the residence time drive the reactions pathways that lead to the formation of lighter intermediates that forms the BC1, as already shown in other works [30].

Figure 6 depicts the Pareto chart reporting the absolute standardized effect of the factors for the BC2 yield. In this case, beyond temperature and time that remains important in the generation of BC2 heavy branched molecules, the combined interaction between temperature and B/W becomes significant. This means that a relative variation in the solid material introduced in the slurry, combined with a variation of the reaction temperature, has more influence in the reaction mechanisms that produce BC2 rather than varying the B/W alone.

Interestingly, concerning the total biocrude yield (Figure 7), the most significant factor is the interaction between temperature and B/W, followed by temperature. In this case, time is not significant, suggesting that, in order to detect a statistically significant effect, longer residence time should be investigated. This effect demonstrates that increasing (or decreasing) the biomass content together with a variation in the reaction temperature drive the degradation reactions that form the biocrude (e.g., phenols, methoxyphenols and longer oxygenated aromatics chains). In general, this means that, if the objective is to optimize the process in terms of total biocrude yield without considering its quality, both of these factors have to be jointly taken into account.

In addition to the experiments of the DOE, four other reaction conditions were tested in duplicates, by increasing temperature to 370 °C and time to 15 and 20 min, collecting the products with Procedure 1. Figure 8 reports the solid residue and biocrudes yields from the experiments carried out at a B/W of 10% *w*/*w*. At 370 °C, 5 min, 10% an increase in the yield of the light biocrude and a decrease in that of the heavy one is achieved; BC2 yield decreases with residence time, while BC1 yield reaches the maximum value of 41.7% *w*/*w* at 15 min.

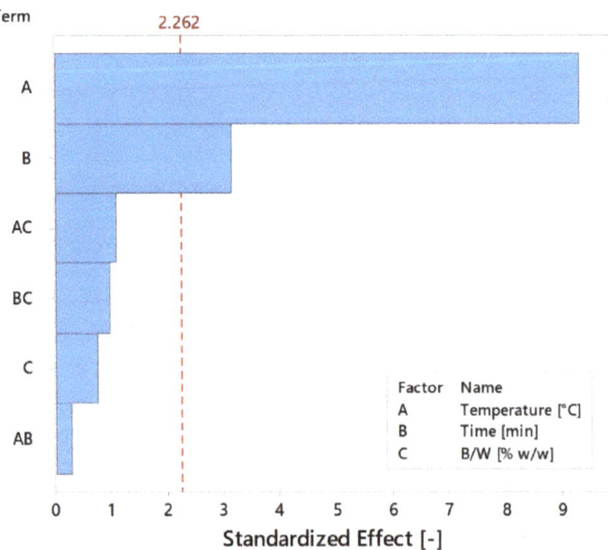

Figure 5. A normal plot of the standardized effects for BC1 yield.

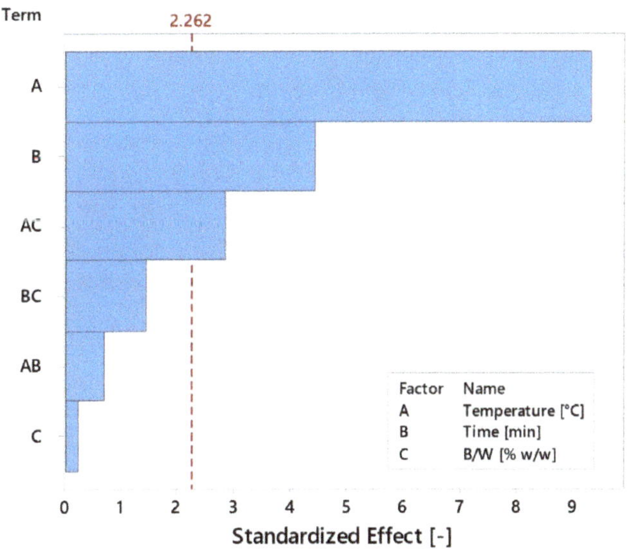

Figure 6. Normal plot of the standardized effects for BC2 yield.

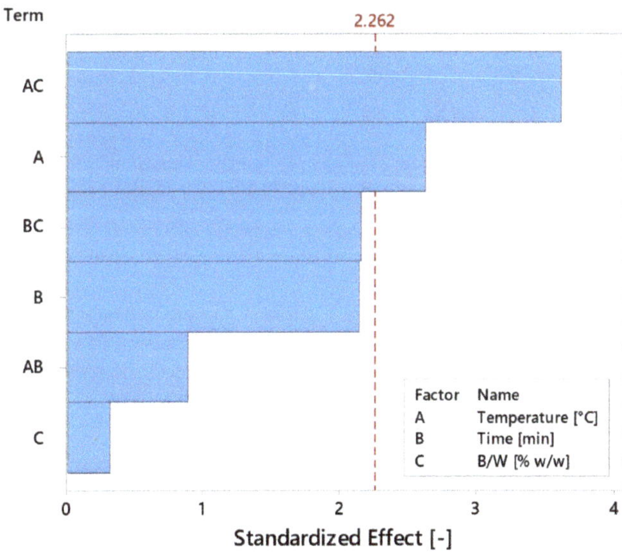

Figure 7. Normal plot of the standardized effects for total biocrude yield.

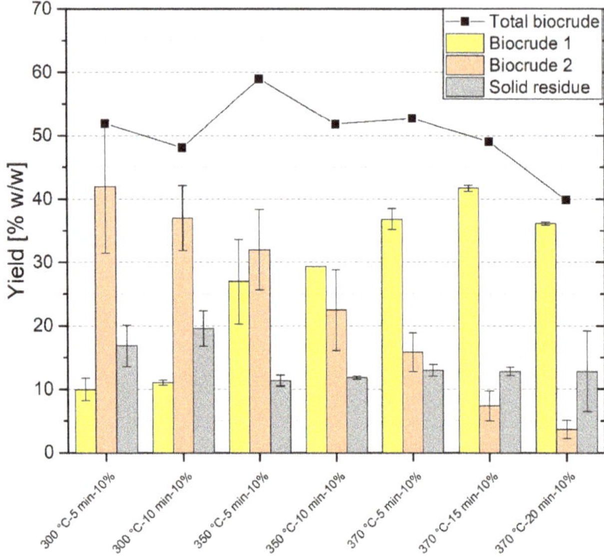

Figure 8. Effect of increased temperature and reaction time on products yield; error bars represent absolute standard deviation.

Although Castello, Pedersen and Rosendahl [9] recently reported that favorable HTL conditions can be obtained also at supercritical condition, it is known that the HTL temperature range where the biocrude is maximized lies between 300 and 350 °C [8,34]. At lower temperatures, partial conversion occurs, whereas at higher values the production shifts towards gases and char. In the present study, the maximum total biocrude yield was obtained at 350 °C, but the peak of its light fraction was achieved at 370 °C, indicating that higher temperatures are needed in order to optimize the conversion of this particular lignin-rich material.

3.4. Elemental Analysis and Higher Heating Value

With respect to the elemental analysis of the LRS (Table 2), both light and heavy biocrude reported an increase in the C and H content and a decrease in O and ash concentration, confirming the energy densification effect of the process. The as-received elemental analysis of biocrude 1 and 2 is reported in Appendix D (Tables A3 and A4). In general, a lower C content and a higher H and O content characterize the light biocrude fraction. These values are in line with literature: Arturi et al. [30] performed batch HTL of Kraft lignin at 300 °C, 15 min, 6% lignin concentration with the addition of 1.6% of K_2CO_3 and obtained a biocrude with 69.9% and 23.6% w/w (d.b.) of carbon and oxygen content, respectively. The feedstock, the two biocrudes and the solid residues CHO compositions are given in the van Krevelen diagram of Figure 9. The light biocrudes have a wider range of H/C and O/C molar ratios with changing reaction conditions, while the heavy biocrudes are less dispersed, having an H/C comprised between 1.10 and 1.25 and an O/C between 0.25 and 0.30. The solid residues are the products that mostly differ from the LRS, having O/C ratios similar to BC2 but lower H/C. The H/C and O/C values of the BC2 obtained in this study are in line with those reported in the review of Ramirez, Brown and Rainey [35], concerning HTL of lignocellulosic biomass. The decrease in the O/C and the increase in the H/C ratio of BC1 with respect to the feedstock suggest that the production of light biocrude was mainly due to decarboxylation rather than dehydration, which, on the contrary, was more evident for the production of BC2 and the solid residues.

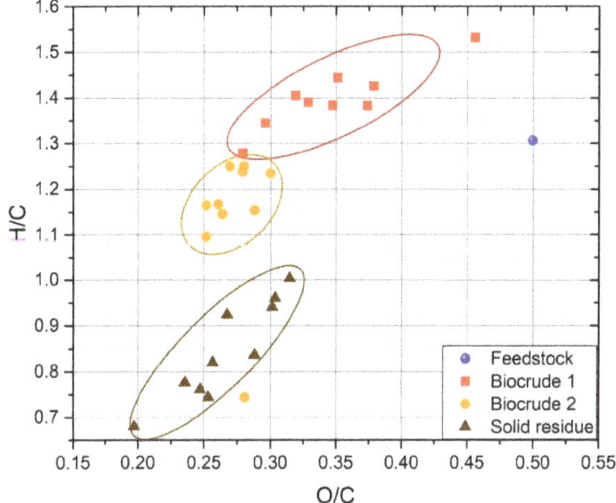

Figure 9. Van Krevelen diagram of lignin-rich stream (feedstock), BC1, BC2 and solid residue (char).

A significant energy densification effect was achieved through the HTL treatment: the higher heating values of the biocrudes ranged between 24.9 and 29.5 MJ kg^{-1} (see Table A6 in Appendix D). Although rather similar values were observed, HHVs of heavy biocrudes were generally higher than those of light ones. The HHV of the total biocrude was determined as a yield-based weight-average from that of BC1 and BC2. The maximum increase with respect to the feedstock (27%) was achieved at 350 °C, 5 min, 10%, the same operating condition, which produced the maximum amount of total biocrude at a B/W of 10%. When BC1 and BC2 are considered separately, their yields and energy densifications, in terms of calorific value, have contrasting trends. Indeed, the yield of BC1 increases with severity, while that of BC2 decreases. The opposite is shown for the HHV: that of BC1 nearly decreases, while that of BC2 increases with severity (Figure 10). However, considering the total biocrude as the sum of BC1 and BC2, both yield and HHV reach a maximum at the same condition, i.e., 350 °C, 5 min, 10%, as the HHV of total biocrude is evaluated as a yield-based weight-average from that of BC1 and BC2.

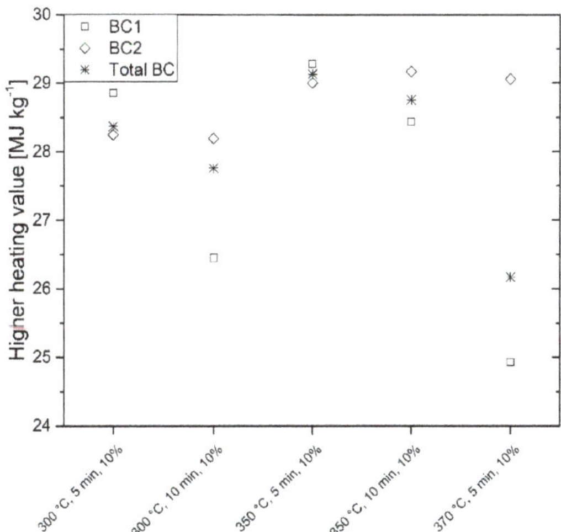

Figure 10. Higher heating value of light, heavy and total biocrude from the experiments carried out at 10% biomass-to-water mass ratio.

3.5. Carbon Balance

Figure 11 shows the carbon balance from the experiments set in the DOE and those carried out at 370 °C, 5 min, 10% and 20%. The balance was reasonably close, ranging from 83% to 108%, with an average value of 92%. The reasons behind this slight underestimated, or in one case, overestimated closure has to be addressed to the approximation of gas composition and to the propagation of errors through products collection and analysis. The majority of the carbon from the lignin-rich stream is retained in the biocrude (from 53.8% to 77.6%). At a low temperature, in general, it is mainly recovered in the heavy biocrude (from 40.9% to 62.3%), while at higher values, especially at 370 °C, it is largely retained in BC1. The carbon ending in the solid residue is not negligible, ranging from 13.0% to 23.6%, indicating that this product can still represent a valuable resource. The percentage of the feedstock carbon retained in the gas phase is the lowest among all products, which is estimated to lay between 1.0% and 3.1%, while a percentage range of 4.6%–11.3% is trapped in the aqueous phase in form of WSO.

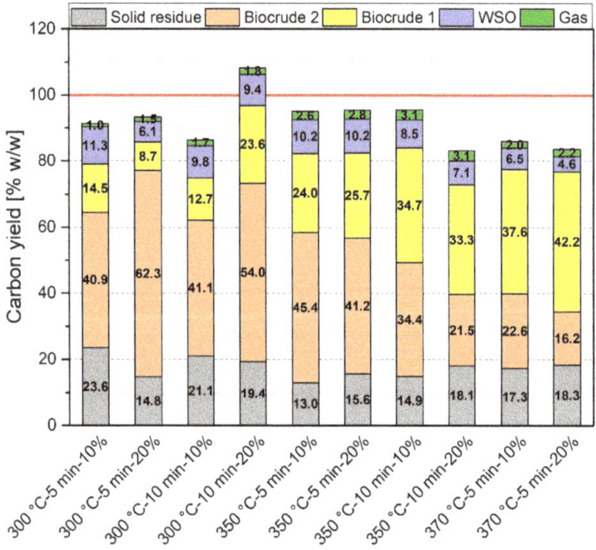

Figure 11. Carbon distribution among HTL products.

3.6. FTIR Analysis

FTIR spectra of the LRS, BC1 and BC2 were qualitatively analyzed in order to evaluate functional moieties modifications after the HTL treatment (Figure 12). If compared with the feedstock, biocrudes exhibit a decrease in intensity at 1000–1070 cm^{-1}, along with the loss of a peak at 1056 cm^{-1}. This is probably due to breaking the β–O–4 or/and α–O–4 ether bonds of lignin, as also confirmed by other studies [20,36,37], suggesting that the feedstock underwent to hydrolysis depolymerization [10,20]. The macromolecular lignin backbone is in fact preferentially fragmented by ether bonds, which can be more easily broken than the C-C linkages through hydrolysis reactions [10]. The two biocrudes show only slight differences. The presence of a relevant peak in the BC2 spectrum around 1700 cm^{-1}, typical of νC=O [20,36–39], is presumably related to the presence of acetone residues, confirmed by the peaks around 1360 and 1419 cm^{-1}. The peaks in the region of aromatics, typical of the lignin structure, around 1600, 1515 and 1460 cm^{-1} [36,37], are always present in the three samples, suggesting that the lignin aromatic rings were, in general, preserved during HTL. An enhancement of the intensity around 1250–1200 cm^{-1} is probably related to guaiacols and mainly syringols [40], first products of lignin depolymerization [10].

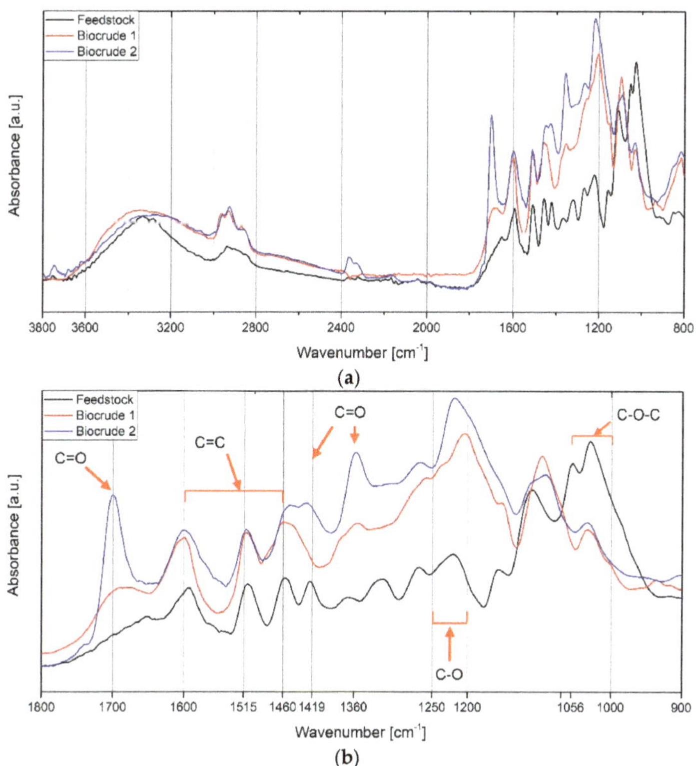

Figure 12. Infrared spectra of the LRS and light and heavy fraction of biocrude from the experiment carried out at 370 °C, 5 min, 20%: (**a**) entire spectra; (**b**) particular.

3.7. Molecular Weight Analysis

In order to gain further insight on differences between the light and the heavy biocrude, their molecular weight (or molar mass) was evaluated by gel permeation chromatography (GPC). The weight-average molecular weight (M_w), the number-average molecular weight (M_n) and the polydispersity index (PDI = M_w/M_n) are reported in Appendix E (Table A7). In addition, it was attempted to determine the average molecular weight of the lignin-rich stream, but only ~10% of this was soluble in THF (ambient temperature) and therefore this value was not estimated. Differently, the biocrude samples were completely THF-soluble and, as expected, the molar masses of the light biocrudes were far lower than the ones of the BC2. The former is comprised between 390 and 490 g mol^{-1}, while the latter range between 1030 and 1400 g mol^{-1}. The values of M_w at 10% and 20% B/W for BC1, BC2 and total biocrude are shown in Figure 13. The M_w of the total biocrude was determined as a yield-based weight-average from that of BC1 and BC2. Concerning BC1, a higher B/W, in general, produces a higher molar mass. An increase with residence time is shown at 300 °C, while an opposite behavior is reported at 350 °C. A maximum is reached at 350 °C, 5 min, both at 10% and 20% *w*/*w* of B/W, though the latter is subjected to high standard deviation, and the minimum values are reached at 370 °C (400 and 391 g mol^{-1}, respectively). Despite the complex trend of the molecular weight of BC2, the M_w of the total biocrude clearly decreases with temperature and time, changing from 1146 to 565 g mol^{-1}, indicating that a higher extent of depolymerization occurred at harsher reaction conditions.

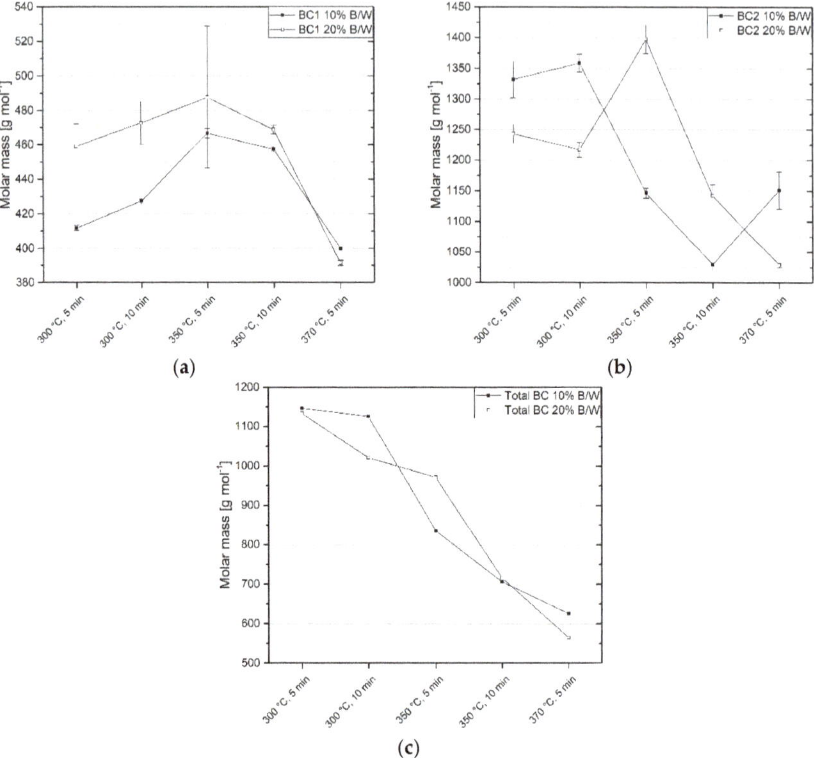

Figure 13. Weight-average molecular weight at different reaction conditions: (**a**) biocrude 1, (**b**) biocrude 2, (**c**) total biocrude. Error bars represent the absolute standard deviation.

Furthermore, it is interesting to look at the effect of reaction conditions on the molar mass distributions (Figure 14). The shape of the distribution is only slightly altered by changing residence time at fixed temperature and B/W but greatly changes with temperature, suggesting the latter to be a more influencing parameter, at least at the investigated reaction conditions. Concerning the light biocrude, at 300 °C and 10%, there is a contribution of low weight compounds (between 40 and 100 g mol^{-1}), which disappears at 350 °C and reappears, even with a slightly different shape, at 370 °C, confirming the presence of a maximum. On the contrary, the molar masses of the heavy biocrudes are more homogeneous, as confirmed by the lower PDI. Considering the BC1 samples obtained at a B/W of 20%, at 300 °C they still exhibit a low-molecular-weight peak, but, in this case, it is narrower and shifted towards higher values, precisely between 100 and 200 g mol^{-1}. The shapes of the distributions of samples at 350 and 370 °C are more similar to the ones of the 10% case. The distributions of BC2 are comparable to those obtained at 10%, only a more marked hump between 500 and 600 g mol^{-1} is present in the samples obtained at 300 °C.

These findings, along with the FTIR results, show that LRS was effectively depolymerized during the HTL treatment, even if without any alkali catalyst or capping agent and with far lower residence time than commonly reported in lignin depolymerization experiments, where the reaction time is generally extended up to several tens of minutes or hours [20,21,29,41].

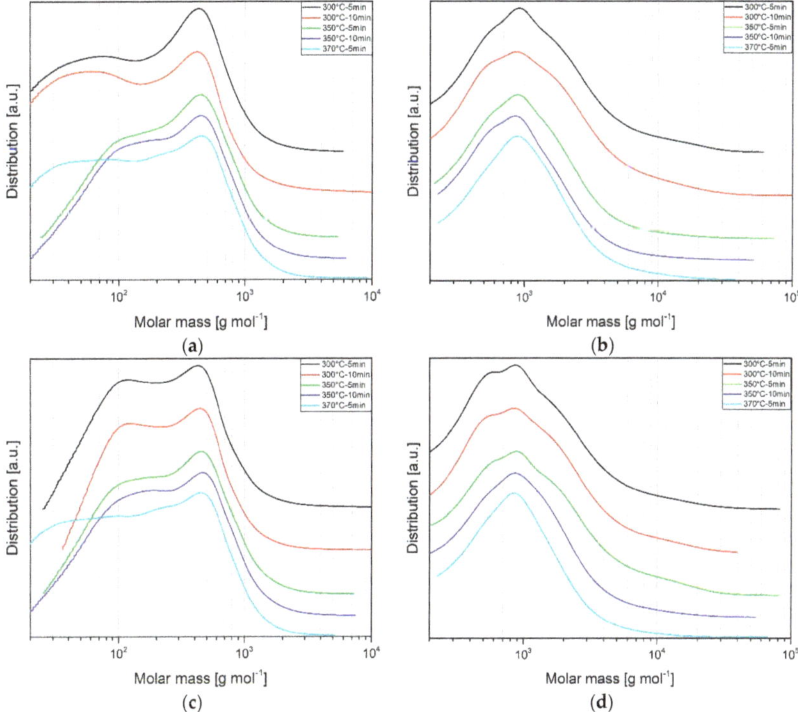

Figure 14. Stacked molecular weight distribution of BC1 (left) and BC2 (right) from experiments carried out at 10% *w*/*w* (**a**,**b**) and 20% *w*/*w* (**c**,**d**).

4. Conclusions

In this work, hydrothermal liquefaction of lignin-rich stream (LRS) from a demo-scale lignocellulosic ethanol plant was investigated without the use of any catalyst or capping agent, recovering two biocrude fractions, a light (BC1) and a heavy one (BC2). Batch lab-scale experiments were carried out and two different collection procedures were developed and compared in terms of yields, biocrude and aqueous phase composition. When performing lab-scale experiments, the use of extraction solvents is somehow mandatory, due to technical limitations related to the small size of reactors, and it was demonstrated here that the collection procedure directly affects yields and products composition. Thus, particular attention should be given when comparing the results from different studies. Indeed, if the aqueous phase is not separated prior to biocrude extraction, a larger amount of biocrude is recovered. On the other hand, removing the process water before biocrude recovery allows for a more suitable comparison with an industrial/continuous process, where the biocrude will be reasonably separated from the water by gravity. Moreover, it was statistically demonstrated that, at the investigated reaction conditions, the most significant factor influencing light, heavy and total biocrudes yield was the reaction temperature. Residence time was significant only as regards the yield of BC1 and BC2, while the biomass-to-water mass ratio (B/W) significantly affected only the BC2 and total biocrude yields by its interaction with temperature. The maximum total biocrude yield (65.7% *w*/*w*) was achieved at 300 °C, 10 min, 20% while the maximum yield of the light fraction (41.7% *w*/*w*) was achieved at 370 °C, 15 min, 10%. These results suggest that the conversion process can be optimized in different ways, depending on the characteristics or on the amount of the biocrude to be obtained. Another interesting result in biorefinery perspective is that the HTL process increased the feedstock energy density up to 27%. The elemental analysis suggests that the light biocrude was mainly produced by decarboxylation reactions rather than dehydration, which was more evident for BC2 and the solid residue. The carbon balance indicated that only a low amount of carbon from the LRS ended up in the aqueous phase as water-soluble organics and its major part was retained in the biocrude (up to 77.6%). At low temperatures, carbon is particularly concentrated in the heavy fraction, while at higher temperatures it moved to the light one (up to 42.2%). The FTIR analysis showed that the lignin aromatic structure was preserved in the two biocrudes, showing that the feedstock was mainly subjected to hydrolysis depolymerization. Indeed, the analysis of the molecular weight confirmed this statement, indicating that a consistent fractionation occurred, especially favored by high temperatures.

The HTL experiments herein reported effectively depolymerized the lignin matrix, preserved the aromatic structure of the feedstock and made available phenolic compounds that are valuable precursors of fuel and chemicals, for further separation, purification and processing. It was therefore shown that LRS has the potential of being a source for valuable chemical intermediates and hydrothermal liquefaction can represent a promising technology for the conversion of this high-moist co-product.

Author Contributions: Conceptualization, A.M.R.; Formal analysis, E.M., S.D. and G.L.; Investigation, E.M., S.D. and G.L.; Methodology, A.M.R. and D.C.; Supervision, D.C.; Validation, A.M.R. and D.C.; Visualization, E.M.; Writing—original draft, E.M.; Writing—review & editing, L.R. and D.C.

Funding: This study was funded by the European Union's Horizon 2020 research and innovation program, project Heat-to-Fuel, under Grant Agreement number 764675.

Acknowledgments: Authors wish to acknowledge Lorenzo Bettucci for his contribution in laboratory analysis and Alberto Bini for his contribution in performing the HTL experiments, TOC analyses, data collection and visualization.

Conflicts of Interest: The authors declare no conflict of interest.

Appendix A

This appendix provides details on which basis the solvent for the extraction of the light biocrude (BC1) was chosen.

The multi-step solvent extraction method was applied by many authors [14,21,30,42–47]. Usually, the first solvent needs to be immiscible in water in order to be easily separated from the aqueous phase. On the other hand, the secondary solvent must dissolve most of the higher molecular weight organic compounds in order to maximize the collections of biocrude. The total yield of biocrude is an over-estimation of the real biocrude obtainable in a scaled-up continuous plant. The maximization of the biocrude yield is necessary on this batch lab-scale bench in order to collect a higher amount of material and facilitate its characterization analysis.

The composition of biocrude obtained from HTL of lignin is known to be mainly composed of aromatic oxygenated compounds (e.g., phenols, methoxyphenols) [30]. Since these classes of compounds have the peculiarity to be polar, the choice of the solvent has fallen on those with similar polarity. The three selected solvents were diethyl ether (DEE), dichloromethane (DCM) and dimethyl ketone (DMK or acetone). The former two were chosen for dissolving the lighter organic compounds since they are polar and slightly miscible in water; while acetone was selected for the heavier fraction dissolution. DEE and DCM have been used for the biocrude extraction in many works, some examples are reported in Reference [14,30,43] and References [44–47] respectively. As extraction solvent for the heavy biocrude, DMK was chosen only on literature guidelines [21,42,46], as BC2 is composed of complex high-molecular aromatic oxygenated polymers, which are not possible to be characterized with the same analytical techniques applied on the light biocrude and due to its low toxicity. THF and methanol are other suitable candidates for BC2 extraction [15,30,42,48,49].

The choice of the extraction solvent was assessed in Procedure 1. The comparison of the GC-detectable compounds in the BC1 extracted with DCM and DEE is reported in Table A1, while the quantitative comparison by GC-FID is shown in Table A2.

The identification and quantification of the different compound classes were carried out in a gas-chromatograph GC-2010 (Shimadzu) equipped with a mass spectrometer GCMS-QP2010 GC 2010 Plus (Shimadzu) and a GC-FID GC 2010 Plus (Shimadzu), both equipped with ZB 5HT Inferno (Zebron) columns (30 m length, internal diameter 0.25 mm, film diameter 0.25 μm). In particular, the GC-MS apparatus was used to investigate the qualitative composition of the sample, comparing the spectrum with a NIST 17 library; GC-FID was used for the quantitative analysis of the selected compounds after a 4-point calibration with pure molecular standards and using o-terphenyl as an internal standard. The analysis was performed with a column flow of 2.02 mL min^{-1} for GC- MS and 3.17 mL min^{-1} in GC- FID, with an initial temperature of 40 °C (holding time 10 min), increased to 200 °C (heating rate 8 °C min^{-1}, holding time 10 min) and then to 280 °C (heating rate 10 °C min^{-1}, holding time 30 min).

Table A1. Comparison of GC-MS identified compounds between DCM and DEE-solubles in BC1 (experiments carried out at 300 °C-10 min-10% with Procedure 1).

Compound Class	DCM-Solubles	DEE-Solubles
Acids	-	Acetic acid
	-	Propionic acid
	-	Isovaleric acid
Ketones	Cyclopentanone	2-Cyclopenten-1-one
	2-Cyclopenten-1-one	2,3-Dimethyl-2-cyclopenten-1-one
	2-Methyl-2-cyclopenten-1-one	3-Ethyl-2-hydroxy-2-cyclopenten-1-one
	3-Ethyl-2-hydroxy-2-cyclopenten-1-one	Acetoin
	Acetoin	Acetovanillone
	Acetosyringone	Acetosyringone
	Desaspidinol	Desaspidinol
Aldehydes	Vanillin	Vanillin
	Syringaldehyde	Syringaldehyde
Phenols	Phenol	Phenol
Methoxyphenols	Syringol	Syringol
	Methoxyeugenol	Methoxyeugenol
	Guaiacol	Guaiacol
	Creosol	Creosol
	4-Propylguaiacol	4-Propylguaiacol
	4-Ethylguaiacol	4-Ethylguaiacol
	-	Eugenol
	-	Isoeugenol

Table A2. Comparison of GC-FID quantification between DCM and DEE-soluble compounds in BC1 (experiment carried out at 300 °C, 10 min, 10% with Procedure 1).

Compound Class	Concentration (μg mL^{-1})	
	DCM-Solubles	DEE-Solubles
Acids	0.28	3.09
Ketones	0.05	0.06
Aldehydes	0.17	0.29
Phenols	3.49	7.28
Methoxyphenols	4.04	7.19
Total	8.03	17.9

The GC-MS qualitative analysis shows that by using DEE as extraction solvent it is possible to identify more compounds with respect to DCM. Moreover, the quantification by GC-FID in the DEE sample gives a total value that is more than double than in the sample collected via DCM. However, with the DCM extraction, the yield of BC1 and BC2 was, respectively 33.4% and 32.8% w/w (d.b.), while in the case of DEE they are 12.3 and 36.3% w/w (d.b.). Dichloromethane is able to extract a higher amount of heavier compounds that are not detectable in GC, leading to a decrease in concentration of the lighter quantifiable organics. After this consideration, DEE was chosen.

Appendix B

This appendix provides details on the effects of temperature, time, biomass-to-water mass ratio and their interactions on BC1, BC2 and total biocrude yield.

Figure A1a shows that, as far as the light biocrude yield is concerned, the temperature is the most influencing parameter, followed by time and B/W: an increase in the value of these parameters leads to an increase in the BC1 yield. Figure A1b shows the influence of the interaction of the process parameters: only at 350 °C and 5 min the effect of B/W was negligible.

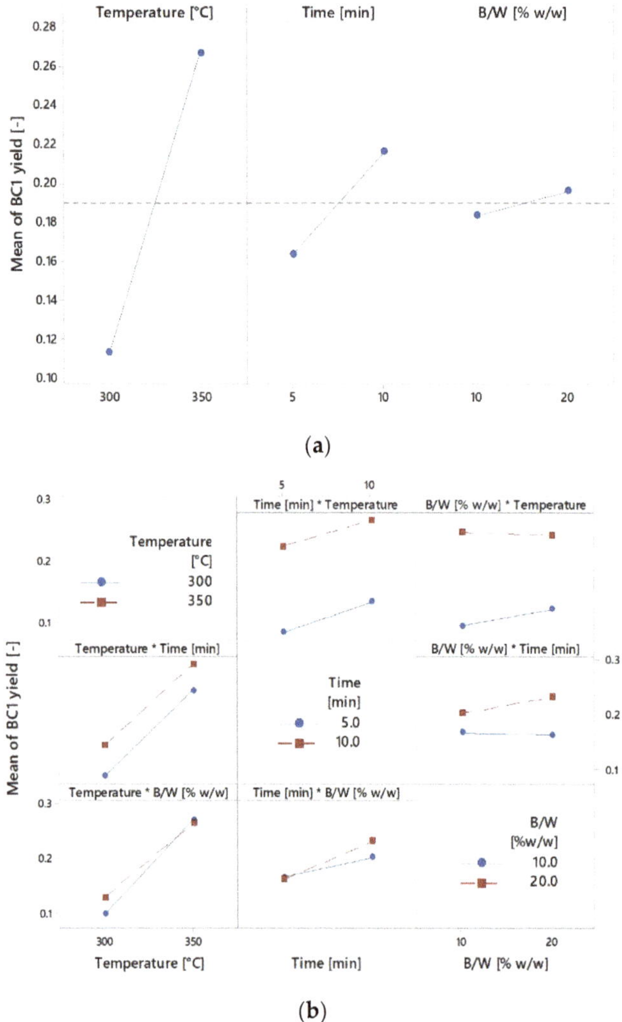

Figure A1. Influence of operating parameters on the yield of biocrude 1: main effects plot (**a**) and interaction effects plot (**b**).

Similarly, Figure A2 reports the influence of the operating parameters and their interactions on the yield of BC2. Again, temperature and time are the most influencing factors, but, in this case, along with their interactions, they negatively affect the yield, except for B/W, when the temperature is 300 °C or the time is 10 min.

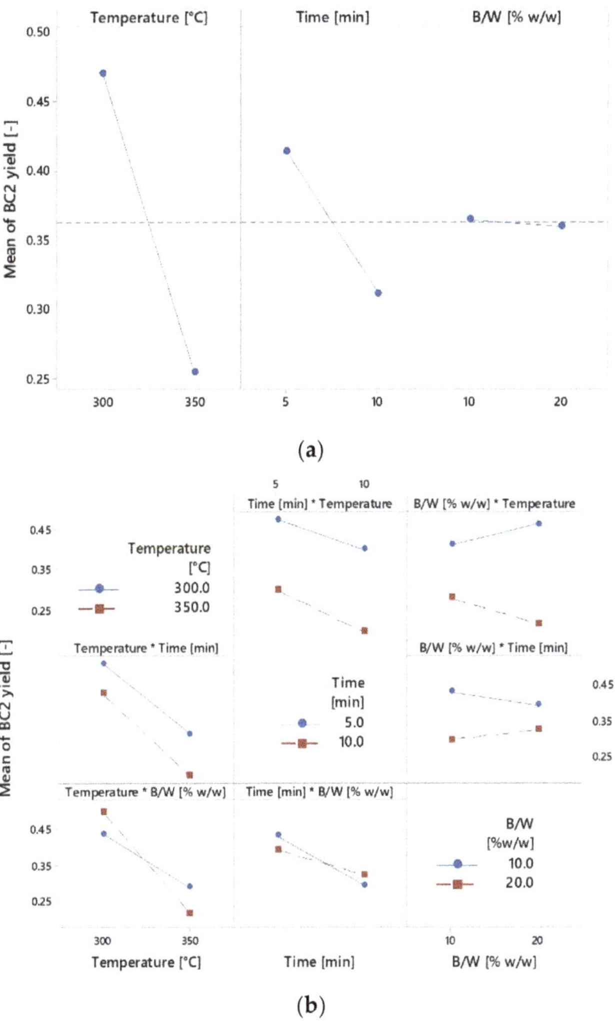

Figure A2. Influence of operating parameters on the yield of biocrude 2: main effects plot (**a**) and interaction effects plot (**b**).

Concerning total biocrude mass yield (Figure A3), an increase in temperature and time leads to a decrease in yield, as for BC2, but B/W imparts an opposite effect, behaving like in the BC1 case. Consequently, the interactions effects are more complex and the total biocrude yield increases with B/W when the temperature is 300 °C or when the residence time is 10 min. It slightly increases with temperature when B/W is 10% w/w and is nearly unaffected by time when B/W is 20% w/w.

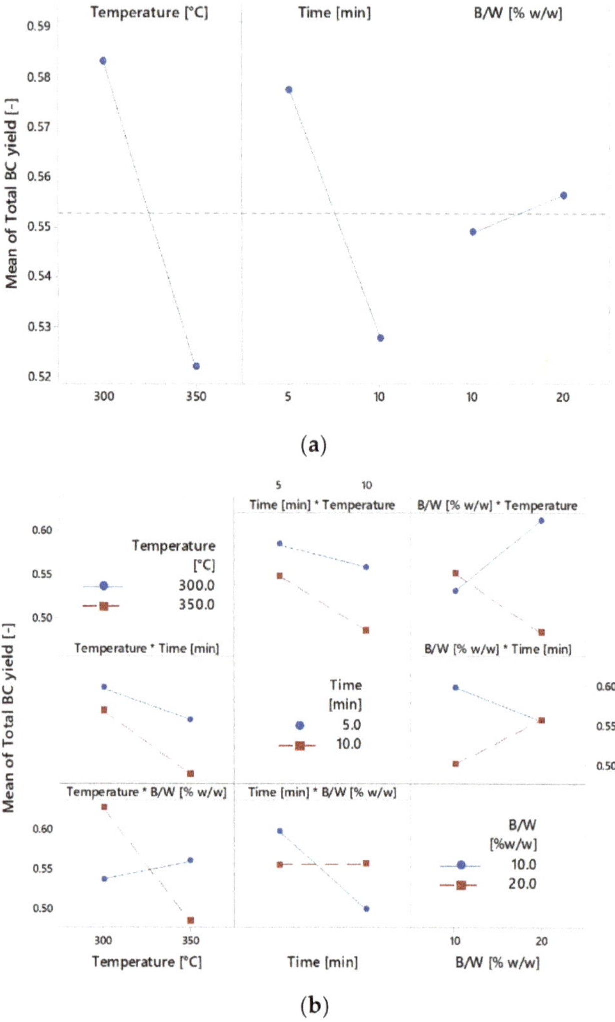

Figure A3. Influence of operating parameters on total biocrude yield: main effects plot (**a**) and interaction effects plot (**b**).

Appendix C

This appendix provides details on the method adopted for the TOC correction, eliminating the contribution of the DEE dissolved in the aqueous samples collected with Procedure 1.

- The TOC and the carbon content of the HPLC-detected compounds (C_{HPLC}) was determined for the aqueous phase of the experiment whose products were collected with Procedure 2 (350 °C, 10 min, 10%) and thus not contaminated with DEE
- The carbon content of the unknown water-soluble species ($C_{unknown}$) was evaluated by the difference
- By supposing the ratio between $C_{unknown}$ and C_{HPLC} to be constant at different HTL conditions and collection procedures, the carbon content due to DEE in the aqueous phase from experiments whose products were collected with Procedure 1 (C_{DEE}) was evaluated according to Equation (A1)

$$C_{DEE} = TOC - C_{HPLC}\left[1 + \left(\frac{C_{unknown}}{C_{HPLC}}\right)_{Proc.2}\right] \quad \text{(A1)}$$

- The corrected TOC was determined by subtracting C_{DEE}

Appendix D

This appendix provides details on the biocrudes' elemental analysis and on the validation of the Channiwala and Parikh equation for the evaluation of the HHV of the BC1 and BC2.

Table A3. Elemental composition of light biocrude fraction (% w/w, as received). Absolute standard deviation, where available, is reported in brackets.

Reaction Condition	Biocrude 1				
	C	H	N	O	Ash
300 °C, 5 min, 10%	65.6 (0.3)	7.35 (0.03)	0.692 (0.028)	25.9 (n.d.)	0.501 (n.d.)
300 °C, 5 min, 20%	61.4 (n.d.)	7.07 (n.d.)	0.687 (n.d.)	30.6 (n.d.)	0.230 (n.d.)
300 °C, 10 min, 10%	60.6 (2.5)	7.20 (0.15)	1.11 (0.33)	30.6 (n.d.)	0.501 (n.d.)
300° C, 10 min, 20%	62.3 (n.d.)	7.49 (n.d.)	0.830 (n.d.)	29.1 (n.d.)	0.287 (n.d.)
350 °C, 5 min, 10%	67.2 (0.6)	7.15 (0.04)	0.683 (0.002)	25.0 (n.d.)	(n.d.)
350 °C, 5 min, 20%	63.9 (0.2)	7.40 (0.04)	0.659 (0.015)	28.0 (n.d.)	0.0108 (n.d.)
350 °C, 10 min, 10%	64.2 (0.4)	7.51 (0.07)	0.930 (0.031)	27.3 (n.d.)	0.0433 (n.d.)
350 °C, 10 min, 20%	62.9 (0.1)	7.25 (0.02)	0.719 (0.008)	29.1 (n.d.)	0.00113 (n.d.)
370 °C, 5 min, 10%	57.1 (1.0)	7.29 (0.04)	0.868 (0.073)	34.7 (n.d.)	0.0160 (n.d.)
370 °C, 5 min, 20%	64.2 (1.2)	7.51 (0.03)	0.930 (0.018)	27.3 (n.d.)	0.0143 (n.d.)

Table A4. Elemental composition of heavy biocrude fraction (% w/w, as received). Absolute standard deviation, where available, is reported in brackets.

Reaction Condition	Biocrude 2				
	C	H	N	O	Ash
300 °C, 5 min, 10%	65.9 (0.6)	6.77 (0.09)	0.847 (0.043)	26.4 (n.d.)	0.136 (n.d.)
300 °C, 5 min, 20%	66.7 (0.1)	6.42 (0.01)	0.948 (0.013)	25.6 (n.d.)	0.146 (n.d.)
300 °C, 10 min, 10%	67.1 (0.2)	6.91 (0.02)	1.13 (0.03)	24.9 (n.d.)	0.0760 (n.d.)
300° C, 10 min, 20%	67.0 (n.d.)	6.98 (n.d.)	0.860 (n.d.)	25.0 (n.d.)	0.138 (n.d.)
350 °C, 5 min, 10%	68.3 (0.4)	6.51 (0.01)	1.07 (0.03)	24.0 (n.d.)	0.124 (n.d.)
350 °C, 5 min, 20%	69.2 (0.7)	6.32 (0.10)	1.13 (0.03)	23.2 (n.d.)	0.122 (n.d.)
350 °C, 10 min, 10%	68.9 (0.4)	6.69 (0.06)	1.110 (0.001)	23.1 (n.d.)	0.232 (n.d.)
350 °C, 10 min, 20%	68.2 (0.3)	4.23 (0.04)	1.82 (0.03)	25.5 (n.d.)	0.240 (n.d.)
370 °C, 5 min, 10%	68.0 (0.5)	6.61 (0.018)	1.28 (0.02)	23.7 (n.d.)	0.457 (n.d.)
370 °C, 5 min, 20%	67.1 (0.2)	6.99 (0.15)	1.278 (0.001)	24.1 (n.d.)	0.470 (n.d.)

Because of the low amount of material produced in each experiment, the higher heating value of the BC1 and BC2 was determined by the Channiwala and Parikh unified correlation [22]:

$$\text{HHV} = 0.3491\text{C} + 1.1783\text{H} + 0.1005\text{S} - 0.1034\text{O} - 0.0151\text{N} - 0.0211\text{Ash}, \quad \text{(A2)}$$

where C, H, O, N, S and Ash respectively represents carbon, hydrogen, oxygen, nitrogen, sulphur and ash content of the sample expressed in mass percentages on dry basis. In the present evaluation, the sulphur content-term was neglected as the LRS has a very limited amount of S (0.2% w/w, d.b.). In order to validate Equation (A2), the HHV of a sample of BC1 and BC2 was measured according to UNI EN 14918 and this value was compared to the Dulong equation [50], which is a correlation widely used in many studies for HHV calculation of HTL biocrude [31,47,51,52]. The results are reported in Table A5: it can be depicted that Equation (A2) leads to a lower relative error (determined according to Equation (A3)).

$$\text{Relative error} = 100 \times |(\text{measured value} - \text{calculated value})| / \text{measured value} \quad (A3)$$

Table A5. Measured and evaluated HHV of BC1 and BC2 samples produced at 350 °C-5 min-20% (Procedure 1) in MJ kg^{-1}.

Value	Biocrude 1	Biocrude 2
Measure	28.47	29.43
Dulong	27.17	28.72
Dulong relative error	4.5%	2.4%
Channiwala and Parikh	28.13	29.53
Channiwala and Parikh relative error	1.2%	0.3%

The HHV values calculated with A2 correlation for BC1, BC2 as well as the yield-based weight-average total biocrude are reported in Table A6 below:

Table A6. Higher heating value of the biocrude samples.

Reaction Condition	Higher Heating Value (MJ kg^{-1})		
	Biocrude 1	Biocrude 2	Total Biocrude
300 °C, 5 min, 10%	28.9	28.3	28.4
300 °C, 5 min, 20%	26.6	29.0	28.6
300 °C, 10 min, 10%	26.5	28.2	27.8
300° C, 10 min, 20%	27.5	29.0	28.6
350 °C, 5 min, 10%	29.3	29.0	29.1
350 °C, 5 min, 20%	28.1	29.5	28.9
350 °C, 10 min, 10%	28.4	29.2	28.7
350 °C, 10 min, 20%	27.5	26.1	27.0
370 °C, 5 min, 10%	24.9	29.1	26.2
370 °C, 5 min, 20%	28.4	29.2	28.6

Appendix E

In this appendix the GPC data are listed in terms of weight-average molecular weight (M_w), number-average molecular weight (M_n) and the polydispersity index (PDI = M_w/M_n) for BC1 and BC2 samples.

Table A7. Weight-average, number-average molar mass (in g mol^{-1}) and a polydispersity index of BC1 and BC2; the absolute standard deviation is reported in brackets.

Reaction condition	Biocrude 1			Biocrude 2		
	M_w	M_n	PDI	M_w	M_n	PDI
300 °C, 5 min, 10%	412 (2)	128 (1)	3.2	1332 (30)	720 (8)	1.8
300 °C, 5 min, 20%	459 (13)	204 (2)	2.2	1243 (16)	668 (9)	1.9
300 °C, 10 min, 10%	427 (2)	101 (1)	4.3	1359 (14)	694 (6)	2.0
300 °C, 10 min, 20%	473 (13)	232 (2)	2.0	1217 (13)	623 (2)	2.0
350 °C, 5 min, 10%	467 (3)	223 (5)	2.1	1147 (9)	697 (10)	1.6
350 °C, 5 min, 20%	488 (41)	265 (76)	1.9	1397 (23)	714 (10)	2.0
350 °C, 10 min, 10%	457 (2)	213 (2)	2.1	1030 (15)	657 (20)	1.6
350 °C, 10 min, 20%	469 (3)	222 (4)	2.1	1142 (18)	701 (9)	1.6
370 °C, 5 min, 10%	400 (0)	105 (1)	3.8	1151 (30)	714 (8)	1.6
370 °C, 5 min, 20%	391 (2)	107 (4)	3.7	1029 (4)	670 (27)	1.5

References

1. European Parliament Directive 2015/1513 of the European Parliament and of the Council of 9 September 2015 amending Directive 98/70/EC relating to the quality of petrol and diesel fuels and amending Directive 2009/28/EC on the promotion of the use of energy from renewable sou. *Off. J. Eur. Union* **2015**, *L239*, 1–29.

2. Bioenergy 2020+ GmbH IEA Task 39 Demo Plant Database. Available online: https://demoplants.bioenergy2020.eu/ (accessed on 12 December 2018).
3. Balan, V.; Chiaramonti, D.; Kumar, S. Review of US and EU initiatives toward development, demonstration, and commercialization of lignocellulosic biofuels. *Biofuels Bioprod. Biorefining* **2013**, *7*, 732–759. [CrossRef]
4. Farag, S.; Chaouki, J. Economics evaluation for on-site pyrolysis of kraft lignin to value-added chemicals. *Bioresour. Technol.* **2015**, *175*, 254–261. [CrossRef]
5. Obydenkova, S.V.; Kouris, P.D.; Hensen, E.J.M.; Heeres, H.J.; Boot, M.D. Environmental economics of lignin derived transport fuels. *Bioresour. Technol.* **2017**, *243*, 589–599. [CrossRef]
6. Porzio, G.F.; Prussi, M.; Chiaramonti, D.; Pari, L. Modelling lignocellulosic bioethanol from poplar: Estimation of the level of process integration, yield and potential for co-products. *J. Clean. Prod.* **2012**, *34*, 66–75. [CrossRef]
7. Xu, C.; Arancon, R.A.D.; Labidi, J.; Luque, R. Lignin depolymerisation strategies: Towards valuable chemicals and fuels. *Chem. Soc. Rev.* **2014**, *43*, 7485–7500. [CrossRef]
8. Cao, L.; Zhang, C.; Chen, H.; Tsang, D.C.W.; Luo, G.; Zhang, S.; Chen, J. Hydrothermal liquefaction of agricultural and forestry wastes: State-of-the-art review and future prospects. *Bioresour. Technol.* **2017**, *245*, 1184–1193. [CrossRef]
9. Castello, D.; Pedersen, T.; Rosendahl, L. Continuous Hydrothermal Liquefaction of Biomass: A Critical Review. *Energies* **2018**, *11*, 3165. [CrossRef]
10. Barbier, J.; Charon, N.; Dupassieux, N.; Loppinet-Serani, A.; Mahé, L.; Ponthus, J.; Courtiade, M.; Ducrozet, A.; Quoineaud, A.A.; Cansell, F. Hydrothermal conversion of lignin compounds. A detailed study of fragmentation and condensation reaction pathways. *Biomass Bioenergy* **2012**, *46*, 479–491. [CrossRef]
11. Wahyudiono; Kanetake, T.; Sasaki, M.; Goto, M. Decomposition of a lignin model compound under hydrothermal conditions. *Chem. Eng. Technol.* **2007**, *30*, 1113–1122. [CrossRef]
12. Zhang, B.; Huang, H.-J.; Ramaswamy, S. Reaction kinetics of the hydrothermal treatment of lignin. *Appl. Biochem. Biotechnol.* **2008**, *147*, 119–131. [CrossRef]
13. Nguyen, T.D.H.; Maschietti, M.; Åmand, L.E.; Vamling, L.; Olausson, L.; Andersson, S.I.; Theliander, H. The effect of temperature on the catalytic conversion of Kraft lignin using near-critical water. *Bioresour. Technol.* **2014**, *170*, 196–203. [CrossRef]
14. Pińkowska, H.; Wolak, P.; Złocińska, A. Hydrothermal decomposition of alkali lignin in sub- and supercritical water. *Chem. Eng. J.* **2012**, *187*, 410–414. [CrossRef]
15. Saisu, M.; Sato, T.; Watanabe, M.; Adschiri, T.; Arai, K. Conversion of Lignin with Supercritical Water—Phenol Mixtures. *Energy Fuels* **2003**, *17*, 922–928. [CrossRef]
16. Jensen, M.M.; Djajadi, D.T.; Torri, C.; Rasmussen, H.B.; Madsen, R.B.; Venturini, E.; Vassura, I.; Becker, J.; Iversen, B.B.; Meyer, A.S.; et al. Hydrothermal Liquefaction of Enzymatic Hydrolysis Lignin: Biomass Pretreatment Severity Affects Lignin Valorization. *ACS Sustain. Chem. Eng.* **2018**, *6*, 5940–5949. [CrossRef]
17. Miliotti, E.; Casini, D.; Lotti, G.; Pennazzi, S.; Rizzo, A.M.; Chiaramonti, D. Hydrothermal Carbonization of Digestate: Characterization of solid and liquid products. In *TC Biomass*; Gas Technology Institute: Chicago, IL, USA, 2017.
18. Toor, S.S.; Rosendahl, L.; Rudolf, A. Hydrothermal liquefaction of biomass: A review of subcritical water technologies. *Energy* **2011**, *36*, 2328–2342. [CrossRef]
19. Peterson, A.A.; Vogel, F.; Lachance, R.P.; Fröling, M.; Antal, M.J., Jr.; Tester, J.W. Thermochemical biofuel production in hydrothermal media: A review of sub- and supercritical water technologies. *Energy Environ. Sci.* **2008**, *1*, 32–65. [CrossRef]
20. Ahmad, Z.; Mahmood, N.; Yuan, Z.; Paleologou, M.; Xu, C. Effects of Process Parameters on Hydrolytic Treatment of Black Liquor for the Production of Low-Molecular-Weight Depolymerized Kraft Lignin. *Molecules* **2018**, *23*, 2464. [CrossRef]
21. Cheng, S.; Wilks, C.; Yuan, Z.; Leitch, M.; Xu, C. (Charles) Hydrothermal degradation of alkali lignin to bio-phenolic compounds in sub/supercritical ethanol and water–ethanol co-solvent. *Polym. Degrad. Stab.* **2012**, *97*, 839–848. [CrossRef]
22. Channiwala, S.A.; Parikh, P.P. A unified correlation for estimating HHV of solid, liquid and gaseous fuels. *Fuel* **2002**, *81*, 1051–1063. [CrossRef]
23. Sluiter, A.; Ruiz, R.; Scarlata, C.; Sluiter, J.; Templeton, D. *Determination of Extractives in Biomass: Laboratory Analytical Procedure (LAP)*; Issue Date 7/17/2005—42619.pdf. Technical Report NREL/TP-510-42619; NREL: Golden, Colorado, 2008; pp. 1–9.

24. Sluiter, A.; Hames, B.; Ruiz, R.; Scarlata, C.; Sluiter, J.; Templeton, D.; Nrel, D.C. *Determination of Structural Carbohydrates and Lignin in Biomass Determination of Structural Carbohydrates and Lignin in Biomass*; Technical Report NREL/TP-510-42618; NREL: Golden, Colorado, 2011.
25. Sluiter, A.; Hames, B.; Ruiz, R.O.; Scarlata, C.; Sluiter, J.; Templeton, D.; Energy, D.; Dötsch, A.; Severin, J.; Alt, W.; Galinski, E.a.; Kreft, J.-U. *Determination of Ash in Biomass. Technical Report NREL/TP-510-42622*; NREL: Golden, Colorado, 2008.
26. Sluiter, A.; Hames, B.; Ruiz, R.; Scarlata, C.; Sluiter, J.; Templeton, D.; Nrel, D.C. *Determination of Sugars, Byproducts, and Degradation Products in Liquid Fraction Process Samples, Technical Report NREL/TP-510-42623*; NREL: Golden, Colorado, 2008.
27. Kang, S.; Li, X.; Fan, J.; Chang, J. Hydrothermal conversion of lignin: A review. *Renew. Sustain. Energy Rev.* **2013**, *27*, 546–558. [CrossRef]
28. Otromke, M.; White, R.J.; Sauer, J. Hydrothermal Base Catalyzed Depolymerization and Conversion of Technical Lignin—An Introductory Review. *Carbon Resour. Convers.* **2019**, *2*, 59–71. [CrossRef]
29. Toledano, A.; Serrano, L.; Labidi, J. Improving base catalyzed lignin depolymerization by avoiding lignin repolymerization. *Fuel* **2014**, *116*, 617–624. [CrossRef]
30. Arturi, K.R.; Strandgaard, M.; Nielsen, R.P.; Søgaard, E.G.; Maschietti, M. Hydrothermal liquefaction of lignin in near-critical water in a new batch reactor: Influence of phenol and temperature. *J. Supercrit. Fluids* **2017**, *123*, 28–39. [CrossRef]
31. Li, Q.; Liu, D.; Hou, X.; Wu, P.; Song, L.; Yan, Z. Hydro-liquefaction of microcrystalline cellulose, xylan and industrial lignin in different supercritical solvents. *Bioresour. Technol.* **2016**, *219*, 281–288. [CrossRef]
32. Orebom, A.; Verendel, J.J.; Samec, J.S.M. High Yields of Bio Oils from Hydrothermal Processing of Thin Black Liquor without the Use of Catalysts or Capping Agents. *ACS Omega* **2018**, *3*, 6757–6763. [CrossRef]
33. Montgomery, D.C. *Design and Analysis of Experiments*, 5th ed.; Wiley: New York, NY, USA, 1997; ISBN 0471316490.
34. Akhtar, J.; Amin, N.A.S. A review on process conditions for optimum bio-oil yield in hydrothermal liquefaction of biomass. *Renew. Sustain. Energy Rev.* **2011**, *15*, 1615–1624. [CrossRef]
35. Ramirez, J.A.; Brown, R.J.; Rainey, T.J. A review of hydrothermal liquefaction bio-crude properties and prospects for upgrading to transportation fuels. *Energies* **2015**, *8*, 6765–6794. [CrossRef]
36. Jiang, W.; Lyu, G.; Wu, S.; Lucia, L.A. Near-critical water hydrothermal transformation of industrial lignins to high value phenolics. *J. Anal. Appl. Pyrolysis* **2016**, *120*, 297–303. [CrossRef]
37. Nazari, L.; Yuan, Z.; Souzanchi, S.; Ray, M.B.; Xu, C. Hydrothermal liquefaction of woody biomass in hot-compressed water: Catalyst screening and comprehensive characterization of bio-crude oils. *Fuel* **2015**, *162*, 74–83. [CrossRef]
38. Bui, N.Q.; Fongarland, P.; Rataboul, F.; Dartiguelongue, C.; Charon, N.; Vallée, C.; Essayem, N. FTIR as a simple tool to quantify unconverted lignin from chars in biomass liquefaction process: Application to SC ethanol liquefaction of pine wood. *Fuel Process. Technol.* **2015**, *134*, 378–386. [CrossRef]
39. Watanabe, M.; Kanaguri, Y.; Smith, R.L. Hydrothermal separation of lignin from bark of Japanese cedar. *J. Supercrit. Fluids* **2018**, *133*, 696–703. [CrossRef]
40. Zhao, J.; Xiuwen, W.; Hu, J.; Liu, Q.; Shen, D.; Xiao, R. Thermal degradation of softwood lignin and hardwood lignin by TG-FTIR and Py-GC/MS. *Polym. Degrad. Stab.* **2014**, *108*, 133–138. [CrossRef]
41. Okuda, K.; Umetsu, M.; Takami, S.; Adschiri, T. Disassembly of lignin and chemical recovery—Rapid depolymerization of lignin without char formation in water-phenol mixtures. *Fuel Process. Technol.* **2004**, *85*, 803–813. [CrossRef]
42. Yang, X.; Lyu, H.; Chen, K.; Zhu, X.; Zhang, S.; Chen, J. Selective Extraction of Bio-oil from Hydrothermal Liquefaction of Salix psammophila by Organic Solvents with Different Polarities through Multistep Extraction Separation. *BioResources* **2014**, *9*, 5219–5233. [CrossRef]
43. Nguyen Lyckeskog, H.; Mattsson, C.; Åmand, L.E.; Olausson, L.; Andersson, S.I.; Vamling, L.; Theliander, H. Storage Stability of Bio-oils Derived from the Catalytic Conversion of Softwood Kraft Lignin in Subcritical Water. *Energy Fuels* **2016**, *30*, 3097–3106. [CrossRef]
44. Grigoras, I.F.; Stroe, R.E.; Sintamarean, I.M.; Rosendahl, L.A. Effect of biomass pretreatment on the product distribution and composition resulting from the hydrothermal liquefaction of short rotation coppice willow. *Bioresour. Technol.* **2017**, *231*, 116–123. [CrossRef]

45. Villadsen, S.R.; Dithmer, L.; Forsberg, R.; Becker, J.; Rudolf, A.; Iversen, S.B.; Glasius, M. Development and Application of Chemical Analysis Methods for Investigation of Bio-Oils and Aqueous Phase from Hydrothermal Liquefaction of Biomass. *Energy Fuels* **2012**, *26*, 6988–6998. [CrossRef]
46. Cheng, S.; D'cruz, I.; Wang, M.; Leitch, M.; Xu, C. (Charles) Highly Efficient Liquefaction of Woody Biomass in Hot-Compressed Alcohol−Water Co-solvents. *Energy Fuels* **2010**, *24*, 4659–4667. [CrossRef]
47. Duan, P.; Savage, P.E. Hydrothermal Liquefaction of a Microalga with Heterogeneous Catalysts. *Ind. Eng. Chem. Res.* **2011**, *50*, 52–61. [CrossRef]
48. Li, C.; Yang, X.; Zhang, Z.; Zhou, D.; Zhang, L.; Zhang, S.; Chen, J. Hydrothermal Liquefaction of Desert Shrub Salix psammophila to High Value-added Chemicals and Hydrochar with Recycled Processing Water. *BioResources* **2013**, *8*, 2981–2997. [CrossRef]
49. Wahyudiono; Sasaki, M.; Goto, M. Recovery of phenolic compounds through the decomposition of lignin in near and supercritical water. *Chem. Eng. Process. Process Intensif.* **2008**, *47*, 1609–1619. [CrossRef]
50. WA, S.; IH, G. Caloric value of coal. In *Chemistry of Coal Utilization Vol. 1*; Lowry, H., Ed.; Wiley: New York, NY, USA, 1945; p. 139.
51. Biller, P.; Riley, R.; Ross, A.B. Catalytic hydrothermal processing of microalgae: Decomposition and upgrading of lipids. *Bioresour. Technol.* **2011**, *102*, 4841–4848. [CrossRef]
52. Feng, S.; Yuan, Z.; Leitch, M.; Xu, C.C. Hydrothermal liquefaction of barks into bio-crude—Effects of species and ash content/composition. *Fuel* **2014**, *116*, 214–220. [CrossRef]

Article

Catalytic Hydrotreatment of Microalgae Biocrude from Continuous Hydrothermal Liquefaction: Heteroatom Removal and Their Distribution in Distillation Cuts

Muhammad Salman Haider, Daniele Castello, Karol Michal Michalski, Thomas Helmer Pedersen and Lasse Aistrup Rosendahl *

Department of Energy Technology, Aalborg University, Pontoppidanstræde 111, 9220 Aalborg Øst, Denmark; mush@et.aau.dk (M.S.H.); dac@et.aau.dk (D.C.); karol.michal.michalski@gmail.com (K.M.M.); thp@et.aau.dk (T.H.P.)
* Correspondence: lar@et.aau.dk; Tel.: +45-9940-9263

Received: 27 October 2018; Accepted: 26 November 2018; Published: 1 December 2018

Abstract: To obtain drop-in fuel properties from 3rd generation biomass, we herein report the catalytic hydrotreatment of microalgae biocrude, produced from hydrothermal liquefaction (HTL) of *Spirulina*. Our contribution focuses on the effect of temperature, initial H_2 pressure, and residence time on the removal of heteroatoms (O and N) in a batch hydrotreating setup. In contrast to common experimental protocols for hydrotreating at batch scale, we devised a set of two-level factorial experiments and studied the most influential parameters affecting the removal of heteroatoms. It was found that up to 350 °C, the degree of deoxygenation (de-O) is mainly driven by temperature, whereas the degree of denitrogenation (de-N) also relies on initial H_2 pressure and temperature-pressure interaction. Based on this, complete deoxygenation was obtained at mild operating conditions (350 °C), reaching a concurrent 47% denitrogenation. Moreover, three optimized experiments are reported with 100% removal of oxygen. In addition, the analysis by GC-MS and Sim-Dis gives insight to the fuel quality. The distribution of heteroatom N in lower (<340 °C) and higher (>340 °C) fractional cuts is studied by a fractional distillation unit following ASTM D-1160. Final results show that 63–68% of nitrogen is concentrated in higher fractional cuts.

Keywords: hydrothermal liquefaction (HTL); *Spirulina*; hydroprocessing; hydrotreating; upgrading; hydrodeoxygenation (HDO); hydrodenitrogenation (HDN); fractional distillation; drop-in biofuels; nitrogen distribution

---

1. Introduction

Unprecedented climate shift and brisk depletion of conventional resources have raised socioeconomic and environmental concerns; thus the pursuit of clean, independent and alternative renewable and sustainable commercialized solutions for fuels has been expedited [1]. Fuels produced from so-called 2nd and 3rd generation biomass feedstocks, such as sewage sludge and microalgae, offer the advantage of a potentially favorable carbon footprint and a lower risk of negative indirect land use change (ILUC) [2–4].

Microalgae have been identified as an attractive energy source for the production of biofuels due to their ability to accumulate relatively large amount of lipids along with carbohydrates, and proteins [5,6]. Microalgae show higher photosynthetic efficiencies compared to terrestrial plants which could be beneficial in terms of environmental impact [7]. High growth rates and non-competition with arable land and fresh water resources, triggered escalating interests to convert microalgae into liquid fuels, which engage researchers to develop environment-friendly methods with remarkable potential roots for mass production [8,9].

Out of all available energy valorization techniques at our disposal, hydrothermal liquefaction (HTL) is considered as one of the most promising technologies for the production of biocrude oil, due to its advantages in rapid reaction, using all types of wet or dry feedstocks with no lipid-content restriction [10–12]. Recent techno-economic analyses and life cycle assessments on the conversion of microalgae into liquid fuels show that HTL has lower GHG emissions and higher economic potential, with a better energy return on investment as compared to conventional lipid extraction and transesterification technologies [13–16].

At temperatures around 300–420 °C and pressure above the saturation pressure of water the HTL process produces a viscous and energetically dense black liquid usually referred as biocrude. This biocrude contains ~10–15 wt. % of heteroatoms (primarily O and N) with a high heating value (HHV) of ~30–38 MJ/kg [17,18]. Biocrude as such cannot be directly utilized in the transportation sector. To produce fungible on-specification transportation fuel, further upgrading processes such as deoxygenation, denitrogenation, etc. are required. Over the years, researchers have investigated both heterogeneous and homogeneous in-situ catalysts to achieve improved bio-diesel [19,20]. Catalytic upgrading of biocrude through hydroprocessing is the most promising pathway for the conversion of polar compounds into hydrocarbon-rich mixtures [21]. During hydroprocessing, undesirable components such as metals, oxygen, sulfur, nitrogen, olefins, and aromatics are removed from biocrude by selectively reacting them in the presence of a heterogeneous catalyst, at relatively high pressure (30–170 bar H_2) and at high temperature 300–450 °C [22].

Several studies on the hydrotreating of algal biocrudes are documented in the open literature. Biller et al. [21] investigated the hydroprocessing of biocrude obtained from *Chlorella* microalgae, by using sulfided CoMo and NiMo catalysts at 350 °C and 405 °C, with an initial H_2 pressure between 60–66 bar and a residence time of 2 h. Relatively low yields (41–69%) were observed at high temperature, and 86 and 91% reduction of O, along with 60 and 55% of reduction of N, were obtained with NiMo and CoMo catalyst, respectively. Both catalysts showed no difference in their catalytic activity at batch scale; furthermore, after pentane extraction the oil analysis indicated that the majority of remaining O is contained in high molecular weight compounds, while no adequate reduction of nitrogen was achieved. Bai et al. [23] carried out the catalytic hydrothermal processing of biocrude in the presence of 15 different catalysts in a batch reactor. By using a combination of Ru/C and Raney Ni catalyst they were able to obtain 2 wt. % O and 2 wt. % N with a yield of 77%. Duan et al. [24] observed, 90% of O and 59% of N removal after the catalytic hydrothermal upgrading of microalgae biocrude by using a mixture of Ru/C and Mo_2/C as catalyst at 400 °C for 4 h. Li and Savage [25] explored the degree of deoxygenation and denitrogenation by using an acidic cracking catalyst (HZSM-5) in the presence of H_2 at 400, 450 and 500 °C. They found that the upgraded oil yield was reduced from 75 to 42% as the temperature increases. Elliott et al. [26] reported the continuous hydrotreatment of biocrudes from four different microalgae in a bench-scale trickle bed reactor in the presence of sulfided Co-Mo/γ-Al_2O_3 catalyst at 405 °C and operating pressure of 136 bar. They obtained oil yields of 80–85% with final O and N levels of ~1 and ~0.1 wt. %, respectively. However, the authors also highlighted the difficulty to compare results from continuous and batch system, because the products from batch system are equilibrium limited.

Although these studies featured hydrotreating tests in different process conditions, the overall effect of temperature, pressure and residence time on heteroatom removal has not been systematically evaluated. A thorough investigation could be of great importance to understand which parameters, or combinations thereof, influence the performance of the process. In the current work, a two-level factorial experimental design was developed and the HTL hydrotreatment of microalgae biocrude, obtained from *Spirulina* was evaluated. The effect of temperature, H_2 pressure, and residence time on the hydrodeoxygenation (HDO), hydrodenitrogenation (HDN), and hydrogen consumption was discussed. In this evaluation, a maximum temperature of initially 350 °C, then 400 °C was considered, which differs from the previous literature (e.g., [21,23,26]). Following this analysis, three confirmatory experiments were performed at high temperatures to understand the significance of heteroatom removal. Moreover, a small distillation unit (following ASTM D-1160) was used to fractionate the hydrotreated oils showing complete deoxygenation, in order to investigate the concentration of nitrogen and its distribution in lower (<340 °C) and higher (>340 °C) fractional cuts. Comprehensive analysis on the upgraded fuel and distilled fractional cuts was carried out.

2. Materials and Methods

2.1. Materials

The biocrude used in this research was obtained from Aarhus University (Denmark), after the processing of *Spirulina* under sub-critical conditions at 220 bar and 350 °C. In the framework of the HyFlexFuel project a pilot-scale continuous HTL facility was used to produce biocrude with a throughput of up to 100 L/h [17]. Properties of the feedstock and HTL biocrude are listed in Table 1.

Table 1. *Spirulina* feedstock and HTL biocrude analysis. Elemental composition is expressed on dry basis and oxygen by difference.

Materials	Elemental Composition (wt. %)					HHV (MJ/kg)	Ash Content (wt. %)	Water Content (wt. %)
	C	H	N	S	O			
Spirulina	53.5	7.2	12.6	-	26.6	24.0	5.8	6.4
Biocrude	78.1	10.4	8.0	-	3.5	38.0	0.2	3.8

A pre-activated commercial Ni-Mo/γ-Al_2O_3 hydrotreating catalyst in the form of extrudates was provided by Shell Denmark A/S. This pre-sulfided catalyst was activated at 340 °C and 60 bar in the Shell refinery in Fredericia, Denmark.

2.2. Experimental Methods

2.2.1. Design of Experiments

In order to understand the most influential process parameters during hydrotreating, a factorial two-level experimental design with three factors on two levels (2^3), as stated in Table 2, was used to investigate further [27].

Table 2. Test factors for hydrotreating experiments.

Factor	Name	Unit	Low Level (−)	High Level (+)
A	Temperature	°C	250	350
B	H_2 pressure	bar	40	80
C	Residence time	h	2	4

The complete experimental matrix with obtained responses is given in Table 3.

Even though hydrotreating is affected by a large number of factors, during this study only reaction temperature, initial H_2 pressure, and residence time were considered and thus, the oxygen and nitrogen contents were measured as response variables.

Table 3. Two-level factorial experimental matrix (1–8) along with three confirmatory experiments (9–11) at high temperatures.

Exp.	Factor A Temperature	Factor B Initial H_2 Pressure	Factor C Reaction Time
1	250 (−)	40 (−)	2 (−)
2	350 (+)	40 (−)	2 (−)
3	250 (−)	40 (−)	4 (+)
4	350 (+)	40 (−)	4 (+)
5	250 (−)	80 (+)	2 (−)
6	350 (+)	80 (+)	2 (−)
7	250 (−)	80 (+)	4 (+)
8	350 (+)	80 (+)	4 (+)
9	375 (+)	70 (+)	3 (+)
10	400 (+)	65 (+)	2.5 (+)
11	400 (+)	70 (+)	2 (−)

2.2.2. Experimental Set-Up

Hydrotreatment was carried out in a pair of 25 mL Swagelok micro-batch stainless steel reactors. To confirm reproducibility and comparability of results, experiments were carried out in duplicates. Each measured quantity was represented as the mean of the two independent experiments, with an error corresponding to their standard deviation. For each experiment, 4 g of biocrude were loaded inside the reactor, along with 2 g of pre-sulfided catalyst and three stainless steel spheres (4 mm dia.) to enhance mixing. Both reactors were purged with N_2 and H_2 twice at 80 bar, leak tested and filled with a certain H_2 pressure. In order to obtain high temperatures and efficient mixing a SBL-2D fluidized sand-bath (Techne, Stone, UK) and an agitation device with a frequency of 450 min^{-1} was used. Pressure was continuously measured with an A-10 pressure transducer (Wika, Klingenberg, Germany) and recorded in a data logger connected to a LabVIEW™ programme. After the desired reaction time, both reactors were quenched in a water bath prior to gas venting and product separation. The gas was released in the fume-hood and the hydrotreated liquid products were separated into oil phase and water phase, by centrifugation (6-16HS centrifuge, Sigma, Wem, UK) at 4000 rpm for 5 min.

2.2.3. Fractional Distillation

Selected samples of the produced oil underwent fractional distillation in a small unit, complying with ASTM D-1160 [28]. The set-up was bought from Ace Glass Inc. (Vineland, NJ, USA), and involves a 10 mL pot flask at the bottom of the distillation column along with three graduated distillation receiver tubes having a volume of 3 mL each. This unit is able to reach an operating pressure of 10 mbar after connecting the vacuum pump to the top of the condenser. Water circulation in the distillation column was done at 5 °C by using a 200 F thermal circulation bath (Julabo, Seelbach, Germany) in order to achieve better condensation and to reduce losses in the cold trap. To provide adiabatic conditions and avoid heat losses the column was surrounded by a HS-450 °C NiCr heating cable from Horst GmbH (Lorsch, Germany) with a 4 mm winding radius. Furthermore, the heating tape was insulated from outside and K-type thermocouples were used to measure the vapor temperature and the column head skin temperature. To ensure the adiabatic conditions, the temperature of the heating cable has been kept 5–10 °C below vapor temperature. Pressure, vapor temperature, and pot flask skin temperature were continuously monitored, in order to obtain the desired fractional cuts (lower (<340 °C) and higher (>340 °C)).

ASTM D-1160 was used as a guideline to produce fractions with a final cut temperature of 340 °C, which lies up to diesel range [28]. To avoid cracking and thermal degradation of hydrotreated biocrude, due to the presence of possible heteroatoms, the pot flask temperature and skin temperature should not exceed 310 °C [29]. Gasoline cut was attained and collected at <190 °C under atmospheric conditions. Subsequently, the vacuum was lowered to 20 mbar and the fractional cuts at 168 °C and 203 °C vapour temperature, corresponding to atmospheric equivalent temperature (AET) of jet fuel (190–290 °C) and diesel (290–340 °C) were obtained. The observed column head temperature was converted into AET by using a relation described by Maxwell and Bonnel [30].

2.3. Characterization and Analytical Techniques

A 2400 Series II system CHN-O analyzer (ASTM D-5291) from Perkin Elmer (Waltham, MA, USA) with a detection limit of 100 ppm was used to determine the elemental composition of the hydrotreated oil in terms of C, H, and N. Oxygen was calculated by difference. Furthermore, the degree of deoxygenation (de-O) and denitrogenation (de-N) was measured according to Equations (1) and (2):

$$\text{de-O} = \left(1 - \frac{O_{upgr.oil}}{O_{biocrude}}\right) \cdot 100 \quad (1)$$

$$\text{de-N} = \left(1 - \frac{N_{upgr.oil}}{N_{biocrude}}\right) \cdot 100 \quad (2)$$

The higher heating value (HHV) of the biocrudes and the upgraded samples was estimated through the correlation proposed by Channiwala and Parikh [31]:

$$\text{HHV (MJ/kg)} = 0.3491\text{C} + 1.1783\text{H} + 0.1005\text{S} - 0.1034\text{O} - 0.0151\text{N} - 0.0211\text{A} \quad (3)$$

The ideal gas law was used to estimate the H_2 consumption based on initial and final H_2 pressure in the reactor, under the hypothesis of negligible production of other gases:

$$n_{H_2\ consumed} = \frac{P_{initial}\ V}{RT} - \frac{P_{final}\ V}{RT} \quad (4)$$

The boiling range distribution of biocrudes and upgraded oil was determined by means of simulated distillation (Sim-Dis) following ASTM D-7169 [32]. Sim-Dis was equipped with a Zebron ZB-1XT column by Phenomenex (Torrance, CA, USA), and a gas chromatography–flame ionization detector (GC-FID) from Shimadzu Corporation (Kyoto, Japan). To determine the chemical composition, the biocrude and the upgraded samples were analyzed by gas chromatography – mass spectrometry (GC-MS), utilizing a Trace 1300 ISQ QD-Single Quadrupole instrument (Thermo Scientific, Waltham, MA, USA) with a 40 to 300 °C temperature range as described previously [33].

3. Results and Discussion

Hereby, the hydrotreating of *Spirulina* biocrude is conducted as described in Section 2. The results of the single experiment are presented in terms of degree of deoxygenation (de-O), degree of denitrogenation (de-N), and hydrogen consumption as response variables. After the identification of influencing factors, the characterization of upgraded biocrude is discussed in terms of elemental composition, chemical structure and boiling point distribution.

3.1. The Effect of the Process Parameters on Heteroatoms Removal

In order to identify the most influencing parameters during hydrotreating, a two-level factorial design of experiments with varying reaction temperature, initial H_2 pressure, and residence time was performed until experiment 8 as described in Table 3. To check the influence of these varying parameters on bio-oil properties, the effect and the interaction between each parameter have been calculated based on the elemental composition of the upgraded oil as shown in Table 4. Furthermore, the half-normal probability plot is used to assess which factors are important and which are of no significance.

Table 4. Elemental composition (wt. %) along with the obtained responses (de-O and de-N) and their calculated effects until experiment 8. Oxygen is calculated by difference.

Exp.	C	H	N	O	de-O [%]	de-N [%]	H_2 Consumption [kg $_{H2}$/kg $_{feed}$]	H/C [-]	HHV [MJ/kg]
Biocrude	75.01 ± 0.31	10.40 ± 0.09	7.65 ± 0.10	6.94 ± 0.12	-	-	-	1.66 ± 0.02	37.59 ± 0.20
1	76.94 ± 0.49	10.79 ± 0.10	7.19 ± 0.05	5.11 ± 0.61	26	6	0.0025 ± 0.0002	1.68 ± 0.01	38.93 ± 0.21
2	81.24 ± 0.12	11.18 ± 0.03	5.55 ± 0.04	1.95 ± 0.11	72	27	0.0060 ± 0.0001	1.65 ± 0.00	41.24 ± 0.06
3	77.72 ± 0.10	10.76 ± 0.05	6.53 ± 0.05	5.00 ± 0.20	28	15	0.0033 ± 0.0005	1.66 ± 0.01	39.18 ± 0.07
4	82.24 ± 0.22	11.05 ± 0.08	5.44 ± 0.04	1.28 ± 0.35	82	29	0.0058 ± 0.0008	1.61 ± 0.01	41.51 ± 0.14
5	77.16 ± 0.19	10.66 ± 0.02	6.66 ± 0.50	5.53 ± 0.69	24	10	0.0033 ± 0.0002	1.66 ± 0.01	37.59 ± 0.04
6	82.62 ± 0.17	11.87 ± 0.04	4.35 ± 0.12	1.17 ± 0.09	83	43	0.0108 ± 0.0001	1.72 ± 0.00	42.64 ± 0.09
7	76.82 ± 0.03	10.98 ± 0.03	6.36 ± 0.03	6.13 ± 0.03	12	17	0.0045 ± 0.0003	1.67 ± 0.00	38.68 ± 0.04
8	84.31 ± 0.35	12.13 ± 0.06	4.03 ± 0.17	0.00 ± 0.23	100	47	0.0125 ± 0.0002	1.73 ± 0.00	43.70 ± 0.16
9	84.55 ± 0.16	12.66 ± 0.10	3.09 ± 0.05	0.00 ± 0.01	100	60	0.0103 ± 0.0001	1.80 ± 0.00	44.38 ± 0.17
10	84.17 ± 0.02	11.99 ± 0.05	3.84 ± 0.11	0.00 ± 0.04	100	50	0.0080 ± 0.0002	1.71 ± 0.00	43.45 ± 0.06
11	84.37 ± 0.11	12.44 ± 0.02	3.19 ± 0.04	0.00 ± 0.01	100	58	0.0093 ± 0.0002	1.77 ± 0.00	44.05 ± 0.06

Potential coke and product gases during hydrotreatment were considered negligible as compared to the losses during liquid (oil plus water) recovery in a micro-batch reactor. Thus, the overall yield (wt. %) of upgraded oil is not reported in the present study in order not to present an inaccurate picture. Though, the liquid recovered was around ~90 wt. % ± 5 during all experiments which is similar to other works in the literature, for example [21]. Current study focuses on the removal of heteroatoms (O and N) by varying the initial H_2 pressure and reaction time at considerably low operating temperatures (250 °C and 350 °C).

Table 4 shows intriguing results in case of heteroatom removal. Within the detection limit of CHN/O, 100% O content is removed from algal biocrude at 350 °C, 80 bar and 4 h of reaction time. At these severe conditions of present experimental matrix N content is reduced up to 47 %. Till now, the most prominent studies on heteroatom removal by Elliott et al. [26], Biller et al. [21] and Bai et al. [23] showed the removal of O and N up to ~90% and ~99% respectively at ~400 °C.

Biller et al. [21] and Duan et al. [24] reported 91 and 90% reduction of O along with 60 and 59% reduction of N at 405 °C and 400 °C. Furthermore, Elliott et al. [26] was able to reduce final O and N levels down to ~1 and ~0.05 wt. % (99 %) at 405 °C in a continuous flow catalytic fixed bed reactor. Lower denitrogenation rates until experiment 8 from the above study are evidence that high temperatures are vital for the removal of nitrogen. It also seems obvious, that mild conditions relating with each variable factor just slightly affect the O content, whereas the N content almost remain intact.

Based on this knowledge, a set of three confirmatory experiments (Exp. 9, 10 and 11) were performed as shown in Table 4. They are based on the anticipation that N content will be further reduced with the 100 % removal of O at high temperatures. Subsequently, the results presented in Table 4 show complete deoxygenation for experiments 9, 10 and 11, while at best the denitrogenation was only achieved up to 60% at 375 °C and 70 bar initial H_2 for three hours. However, 50% and 58% of denitrogenation is achieved at 400 °C by changing H_2 pressure and reaction time.

The absolute values from Table 5 indicate the most relevant factors affecting the responses (R_1 and R_2). Moreover, half-normal probability graph as shown in Figure 1 is used to assess the significant and insignificant factors for de-O (R_1) and de-N (R_2).

Table 5. Selected process parameters (factors) and their measured effects on de-O (R_1) and de-N (R_2) from experiments 1–8.

Effect	Factor A	Factor B	Factor C	Factor AB	Factor AC	Factor BC	Factor ABC
Effect R_1 [%]	64.24	3.51	6.50	14.63	10.06	0.88	6.03
Effect R_2 [%]	24.07	10.80	4.53	6.29	−1.75	−0.51	1.85

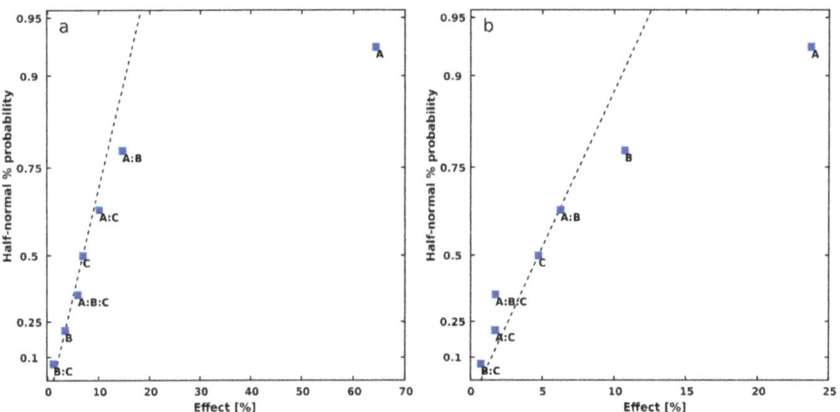

Figure 1. Half normal plot of effects [%], de-O (**a**) and de-N (**b**).

Assuming that the estimated effects of each factor and each interaction term on the de-O follow a normal distribution, the absolute estimated effects will then follow a half-normal distribution. By plotting the ranked estimated effects against the cumulative percent contribution (the contribution to the total sum of squares) of each factor, significant (and insignificant) factors can be identified. Insignificant factors are those having small estimated effects and tend to fall on a straight line starting at the origin (0:0).

From Figure 1a, it is clear that the only significant factor influencing the de-O is the temperature. Obviously, the half-normal plot show only positive effects (absolute estimated effects). By examining Table 4 it is evident that temperature indeed has a positive effect on the de-O. At the lower level temperature, the higher degree of deoxygenation obtained was 28% (Exp. 3), whereas the de-O range from 72–100% at the higher level. Likewise, a similar identification of the significant factors influencing the de-N can be observed in in Figure 1b. It is apparent that temperature is also significantly influencing the de-N along with the hydrogen pressure. Thus, the highest de-N was obtained in Exp. 8 (all factors set at high level) yielding a de-N of 47%, which is only marginally higher than Exp. 6 (de-N = 43%) for which the reaction time was only 2 h. As for the de-O, any interaction effects seem insignificant.

The hydrotreated oil after all 11 experiments has a higher C and a lower O and N content. However, H/C molar ratio remains more or less stable with the algal biocrude at lower temperatures and lower pressures. While, at higher temperatures (350–400 °C) and higher pressures (80 bar) overall H/C molar ratio increases and thus the aromatics in the oil decreases. In addition, a negligible effect of reaction time is observed on the molar H/C ratio. However, at 400 °C lower H/C is observed as compared to 375 °C, which is anticipated as a result of cracking reaction. Subsequently, after the complete deoxygenation and 60% of N reduction the HHV increases from 37.6 MJ/kg of the algal biocrude to 44.4 MJ/kg. Higher temperatures are needed for complete denitrogenation, but both H/C and HHV decrease at higher temperature of 400 °C in relation with 375 °C, which means that we are getting higher degree of denitrogenation with a loss of fuel quality. This may be caused by the formation of coke, as HTL biocrude due to high polarity are unstable and prone toward the polymerization and deactivation of NiMo catalyst. This complete deoxygenated oil with reduced N content is considered more stable and thus reduces the possibility of catalyst deactivation during further upgrading for on-specification fuel.

3.2. Analysis of the Up-Graded Oil

Simulated distillation (Sim-Dis) was used to evaluate the boiling point distribution of the catalytically upgraded oil samples and then compared with the untreated biocrude. Five different boiling point ranges are defined and described as gasoline (>190 °C), jet fuel (190–290 °C), diesel (290–340 °C), vacuum gas oil (340–538 °C) and vacuum residue (>538 °C). Table 6 along with Figure 2 shows yield of different fractions and boiling point distribution of all the experiments with complete degree of deoxygenation along with the biocrude and experiment 1 with the lowest possible operating range (250 °C, 40 bar H_2 for 2 h). The biocrude shows large fractions in high molecular weight compounds and therefore, it associates with the higher residual cuts (72.7% above 340 °C). In addition, the overall recovery at 720 °C increases from 72% to 92%, thus the amount of potential distillable products also increases. Furthermore, after 100% HDO more than 32% and 14% oil is in jet fuel and diesel range and around 15% is confined in residue. High temperatures and high initial H_2 pressures are vital for the cracking of biocrude into lower fractional cuts. Present study shows that at 350 °C and 80 bar initial H_2 pressure highest boiling point distribution above diesel is obtained with minimum residual fraction cut of 12.5%. This looks very promising as by just tuning operating conditions at mild temperature (350 °C) high production of diesel drop-in fuel could be achieved (Table 6 and Figure 2).

Table 6. Sim-Dis analysis of biocrude and selected hydrotreated samples.

Fractional Cuts	Boiling Point Range	Composition [wt. %]					
		Biocrude	Exp. 1	Exp. 8	Exp. 9	Exp. 10	Exp. 11
Gasoline	<190 °C	3.9	3	13.1	4.8	5.6	7.9
Jet fuel	190–290 °C	14.5	15.5	32.2	33.4	34	34.5
Diesel	290–340 °C	8.9	8.5	18.4	15.3	13.6	14.5
Vacuum gas oil	340–538 °C	38.9	34.7	23.8	29	26.3	27.5
Vacuum residue	>538 °C	33.8	38.3	12.5	17.5	20.5	15.6

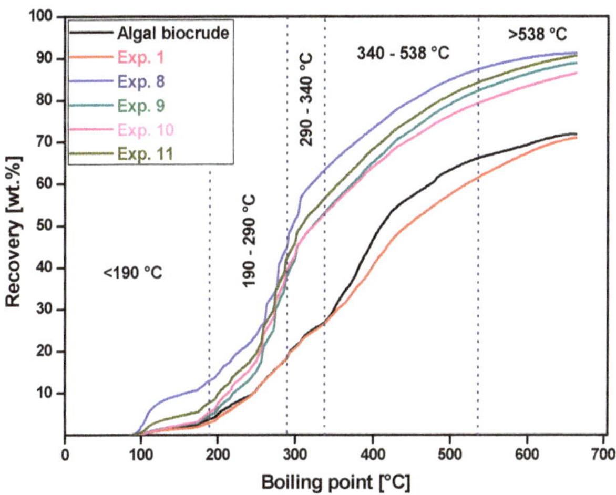

Figure 2. Boiling point distribution of biocrude and upgraded oil samples, obtained by simulated distillation.

Figure 3a,b show the GC-MS chromatograms of the HTL biocrude and the complete deoxygenated oils at 350 °C (47% HDN).

The untreated biocrude (Figure 3a) shows a complex mixture of numerous different hydrocarbons and heteroatoms (mainly O and N), as expected from the elemental analysis. The oxygen containing compounds are mainly saturated fatty acids (hexadecanoic acid) and phenol derivatives such as p- and m-cresol. Moreover, fatty alcohols such as dodecanol are also observed in small amounts. Whereas, nitrogen-containing compounds are mostly found in the form of non-heterocyclic high molecular weight compounds such as various amides. However, N compounds are also present in the form of ring type structures with either one N atom, such as indoles and pyrroles, or two N atoms, such as pyrazines. The sole purpose of hydrotreating is to convert heteroatoms into their respective hydrocarbons along with the saturation of double bonds and cracking of high molecular weight compounds. Figure 3b shows a chromatogram of upgraded oil (Exp. 8) after the complete removal of O atoms at 350 °C and reveals the presence of dominating aliphatic hydrocarbons ranging from C_{14} to C_{21}. Aromatic hydrocarbons such as ethylbenzene and toluene are also present in the upgraded oil. At the same time, N is observed in the GC-MS spectrum, such as the presence of amides in Figure 3b at the retention time of 25.51 min and 27.09 min. However, the GC-MS spectrum also indicates minor presence of amine, indole, nitrile, pyrrole, pyrazine, pyridine and quinoline in the hydrotreated oil.

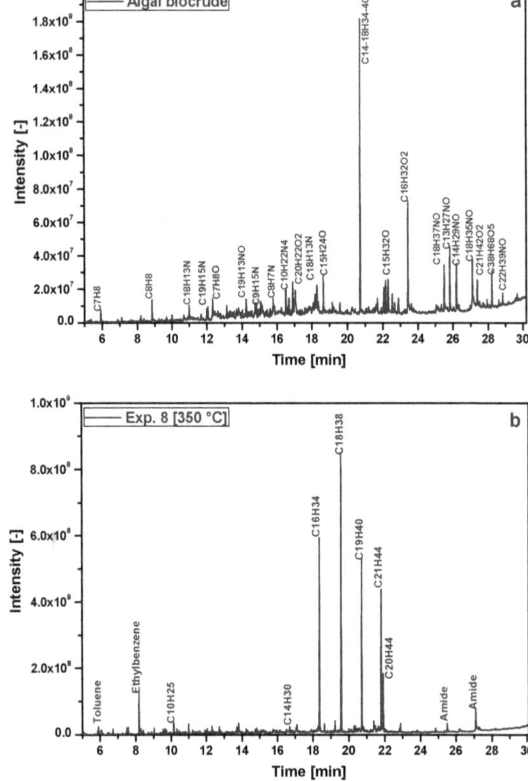

Figure 3. GC-MS chromatographs of (**a**) bio crude and (**b**) Experiment 8 at 350 °C.

3.3. Distribution of Nitrogen-Containing Compounds

GC-MS analysis is only able to investigate the volatile fraction of the sample, hence it does not account for all molecules in the oil. Therefore, in order to assess the distribution of N content in the light and heavy fractional cuts of the upgraded oil, a more relevant industrial approach following ASTM D-1160 is carried out. Biller et al. [21] reported inappreciable amounts of nitrogen-containing compounds through GC-MS; while, through elemental analysis they observed 1–1.5% of O and 2.4–2.7% of N. They performed C_5 (pentane) extraction of hydrotreated oil to investigate the distribution of O and N between C_5 soluble (flowable) and C_5 insoluble (asphaltene) fractions. They were able to completely remove O content in C_5 soluble fraction, but the removal of N content was not satisfactory and it seems that it was almost homogenously distributed in C_5 soluble and insoluble fractions.

In order to determine the distribution of N content in lower and higher molecular weight compounds, the hydrotreated oil from experiments 8 (47% HDN) and 9 (60% HDN) were subjected to further investigation. Consequently, we performed true boiling point distillation following ASTM D-1160 to assess the concentration and the distribution of heteroatom content in low and high molecular weight fuel fractional cuts which correspond up to diesel (<340 °C) and vacuum residue (>340 °C). Results are shown in Table 7.

Table 7. Yields (%) and elemental analysis (N and O content in wt. %), H/C and N/C molar ratios of below and above diesel range fraction cuts.

Exp.	Light Fraction (<340 °C)					Heavy Fraction (>340 °C)				
	Yield [wt. %]	N [wt. %]	O [wt. %]	H/C [-]	N/C [-]	Yield [wt. %]	N [wt. %]	O [wt. %]	H/C [-]	N/C [-]
8	52	1.30	0	1.86	0.02	35	2.73	0.08	1.48	0.03
9	29	1.15	0	1.82	1.32	47	1.95	0.24	1.32	0.02

Table 7 shows, that 32% and 37% of N content is concentrated in the light fraction while vacuum residue contains 68% and 63% of N content for experiments 8 and 9, respectively. These results reveal that most of the nitrogen-containing compounds are distributed in the higher fractional cuts. Additionally, no O content is associated with the light fraction for both experiments while, 0.24 wt. % of O and 0.08 wt. % of O is incorporated in the vacuum residue obtained after the fractional distillation of experiments 8 and 9 respectively. The results indicate that oxygen is in fact concentrated in the heavy fraction, although in small amount. This means that when analyzing the total sample it is diluted to below the error associated with the CHN analyzer and therefore reported as zero. The presence of 0.08 wt. % of O in vacuum residue of hydrotreated oil (375 °C) as compared to the 0.24 wt. % of O (350 °C) also validate the statistical analysis, which showed that temperature is a main driver for O removal. Moreover, overall higher H/C molar ratio is observed for diesel range (>1.82) and very low H/C molar ratio (<1.48) was observed for vacuum residue in case of both experiment.

Figure 4 illustrates a modified Van Krevelen plot as a function of H/C and N/C molar ratio for biocrude, hydrotreating experiments and distillation cut of upgraded oil from experiments 8 and 9. Starting from the same biomass composition, it is clear that both temperature and pressure have a significant effect on the removal of N heteroatoms. Thus, the N/C molar ratio decreases along with the increase of hydrocarbon content. This plot also elucidates the distribution of nitrogen-containing compounds in lower and higher fractional cuts for experiments 8 and 9; and shows that the N/C molar ratio decreases in the hydrocarbon range (higher H/C molar ratio) but on the other hand N/C molar ratio increases as the aromatic content of the fractional cut increases.

Furthermore, the yields of the small fractional distillation unit are also stated in Table 7. 2.76 g and 1.09 g of sample is used for the fractional distillation of experiments 8 and 9, respectively. Higher recovery of 87 wt. % is obtained after the fractional distillation of experiment 8 as compared to 76 wt. % for experiment 9. Vacuum residue cut for both experiments is in line with the Sim-Dis results. However, some losses in the light fractional cut (<340 °C) were observed after comparing it with Sim-Dis results, which may be due to the loss of lighter fraction in the distillation column and column head. From experiment 8 it is clear that the losses are less as compared to experiment 9 due to the use of higher amount of sample. Losses for lighter fractional cuts could be minimized by subjecting higher amount of sample to the fractional distillation unit.

Even though nitrogen-containing compounds are mostly found in the higher fractional cuts (>340 °C), still some nitrogen is found in the light fraction. Therefore, this could not be used directly as a fuel, because of the indirect regulations applied on the fuel specification due to the presence of N compounds. The reason for the presence of remaining N compounds could be found in thermodynamics and the reduction of partial H_2 pressure during hydrotreating. Thus, we suggest that hydrotreating of algal biocrude for the removal of heteroatom N either possibly requires higher temperatures (which will result in reduced yields due to the formation of volatiles) or multi-stage catalytic treatment, to partially stabilize the oil before further upgrading at higher temperatures in order to completely remove nitrogen-containing compounds.

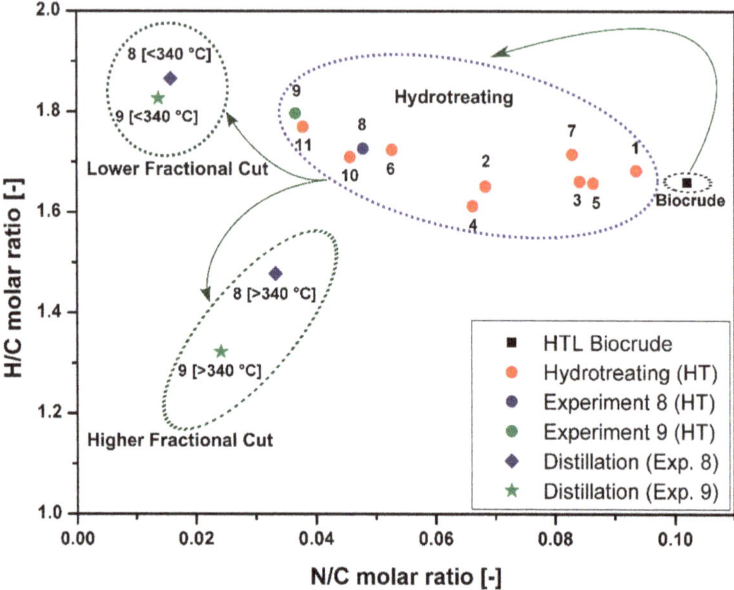

Figure 4. Modified Van Krevelen diagram showing molar ratio of H/C as a function of N/C for *Spirulina* feedstock, HTL biocrude, 11 hydrotreating experiments and fractional distillation cuts of experiment 8 and 9.

4. Conclusions

The most influential parameters affecting the hydrotreating of *Spirulina* (microalgae) biocrude were successfully identified. In addition, the distribution of heteroatoms in lower and higher fractional cuts was explored by using a small fractional distillation unit following the ASTM D-1160. For this purpose, a parametric study following a factorial two-level experimental design with three subsequent confirmatory experiments was carried out in a micro-batch reactor using conventional $NiMo/Al_2O_3$ catalyst. Operating temperature has a key influence on the degree of deoxygenation (de-O), whereas the operating pressure along with temperature-pressure interaction mainly affects the degree of denitrogenation (de-N). Complete deoxygenation of upgraded free flowing oil was observed at 350 °C. An optimized experiment at 375 °C, 70 bar initial H_2 and 3 h residence time leads toward 100% de-O, 60% de-N, 1.80 H/C ratio and a maximum HHV of 44.38 MJ/kg. In addition, fractional distillation up to diesel (<340 °C) and vacuum residue (>340 °C) cut shows a significant concentration of 63–68% of N content in the higher fractional cut. Emanating from these encouraging results, shows that oxygen could be completely removed from given biocrude with a high liquid recovery.

Author Contributions: The authors have equally contributed to the writing of the manuscript.

Funding: This project has received funding from the European Union's Horizon 2020 research and innovation programme under grant agreement No 764734.

Conflicts of Interest: The authors declare no conflict of interests.

References

1. *World Energy Outlook*; International Energy Agency: Paris, France, 2017. Available online: https://webstore.iea.org/world-energy-outlook-2017 (accessed on 14 November 2018).
2. Doornbosch, R.; Steenblik, R. Biofuels: Is the Cure Worse Than the Disease? Available online: https://www.oecd.org/sd-roundtable/papersandpublications/39348696.pdf (accessed on 13 July 2018).

3. Sims, R.E.H.; Mabee, W.; Saddler, J.N.; Taylor, M. An overview of second generation biofuel technologies. *Bioresour. Technol.* **2010**, *101*, 1570–1580. [CrossRef] [PubMed]
4. Alam, F.; Mobin, S.; Chowdhury, H. Third generation biofuel from Algae. *Procedia Eng.* **2015**, *105*, 763–768. [CrossRef]
5. Chisti, Y. Biodiesel from microalgae. *Biotechnol. Adv.* **2007**, *25*, 294–306. [CrossRef] [PubMed]
6. Molino, A.; Iovine, A.; Casella, P.; Mehariya, S.; Chianese, S.; Cerbone, A.; Rimauro, J.; Musmarra, D. Microalgae Characterization for Consolidated and New Application in Human Food, Animal Feed and Nutraceuticals. *Int. J. Environ. Res. Public Health* **2018**, *15*, 2436. [CrossRef] [PubMed]
7. Biller, P.; Ross, A.B. Potential yields and properties of oil from the hydrothermal liquefaction of microalgae with different biochemical content. *Bioresour. Technol.* **2011**, *102*, 215–225. [CrossRef] [PubMed]
8. Singh, N.K.; Dhar, D.W. Microalgae as second generation biofuel. A review. *Agron. Sustain. Dev.* **2011**, *31*, 605–629. [CrossRef]
9. Pienkos, P.T.; Darzins, A. The promise and challenges of microalgal-derived biofuels. *Biofuels Bioprod. Biorefin.* **2009**, *3*, 431–440. [CrossRef]
10. Peterson, A.A.; Vogel, F.; Lachance, R.P.; Fröling, M.; Michael, J.A.; Tester, J.W. Thermochemical biofuel production in hydrothermal media: A review of sub- and supercritical water technologies. *Energy Environ. Sci.* **2008**, *1*, 32–65. [CrossRef]
11. Toor, S.S.; Rosendahl, L.; Rudolf, A. Hydrothermal liquefaction of biomass: A review of subcritical water technologies. *Energy* **2011**, *36*, 2328–2342. [CrossRef]
12. Ramirez, A.J.; Brown, J.R.; Rainey, J.T. A Review of Hydrothermal Liquefaction Bio-Crude Properties and Prospects for Upgrading to Transportation Fuels. *Energies* **2015**, *8*, 6765–6794. [CrossRef]
13. Frank, E.D.; Elgowainy, A.; Han, J.; Wang, Z. Life cycle comparison of hydrothermal liquefaction and lipid extraction pathways to renewable diesel from algae. *Mitig. Adapt. Strateg. Glob. Chang.* **2013**, *18*, 137–158. [CrossRef]
14. Liu, X.; Saydah, B.; Eranki, P.; Colosi, L.M.; Mitchell, B.G.; Rhodes, J.; Clarens, A.F. Pilot-scale data provide enhanced estimates of the life cycle energy and emissions profile of algae biofuels produced via hydrothermal liquefaction. *Bioresour. Technol.* **2013**, *148*, 163–171. [CrossRef] [PubMed]
15. Jones, S.; Zhu, Y.; Anderson, D.; Hallen, R.; Elliott, D.; Schmidt, A.; Albrecth, K.; Hart, T.; Butcher, M.; Drennan, C.; et al. *Process Design and Economics for the Conversion of Algal Biomass to Hydrocarbons: Whole Algae Hydrothermal Liquefaction and Upgrading*; Report No. PNNL-23227; Pacific Northwest National Laboratory: Richland, WA, USA, 2014.
16. Pedersen, T.H.; Hansen, N.H.; Pérez, O.M.; Cabezas, D.E.V.; Rosendahl, L.A. Renewable hydrocarbon fuels from hydrothermal liquefaction: A techno-economic analysis. *Biofuels Bioprod. Biorefin.* **2018**, *12*, 213–223. [CrossRef]
17. Anastasakis, K.; Biller, P.; Madsen, R.; Glasius, M.; Johannsen, I.; Anastasakis, K.; Biller, P.; Madsen, R.B.; Glasius, M.; Johannsen, I. Continuous Hydrothermal Liquefaction of Biomass in a Novel Pilot Plant with Heat Recovery and Hydraulic Oscillation. *Energies* **2018**, *11*, 2695. [CrossRef]
18. Jensen, C.U.; Guerrero, J.K.R.; Karatzos, S.; Olofsson, G.; Iversen, S.B. Fundamentals of Hydrofaction^{TM}: Renewable crude oil from woody biomass, Biomass Convers. *Biorefinery* **2017**, *7*, 495–509.
19. Duan, P.; Savage, P.E. Hydrothermal Liquefaction of a Microalga with Heterogeneous Catalysts. *Ind. Eng. Chem. Res.* **2011**, *50*, 52–61. [CrossRef]
20. Ross, A.B.; Biller, P.; Kubacki, M.L.; Li, H.; Lea-Langton, A.; Jones, J.M. Hydrothermal processing of microalgae using alkali and organic acids. *Fuel* **2010**, *89*, 2234–2243. [CrossRef]
21. Biller, P.; Sharma, B.K.; Kunwar, B.; Ross, A.B. Hydroprocessing of bio-crude from continuous hydrothermal liquefaction of microalgae. *Fuel* **2015**, *159*, 197–205. [CrossRef]
22. Huber, G.W.; Corma, A. Synergies between bio- and oil refineries for the production of fuels from biomass. *Angew. Chem. Int. Ed. Engl.* **2007**, *46*, 7184–7201. [CrossRef] [PubMed]
23. Bai, X.; Duan, P.; Xu, Y.; Zhang, A.; Savage, P.E. Hydrothermal catalytic processing of pretreated algal oil: A catalyst screening study. *Fuel* **2014**, *120*, 141–149. [CrossRef]
24. Duan, P.; Wang, B.; Xu, Y. Catalytic hydrothermal upgrading of crude bio-oils produced from different thermo-chemical conversion routes of microalgae. *Bioresour. Technol.* **2015**, *186*, 58–66. [CrossRef] [PubMed]
25. Li, Z.; Savage, P.E. Feedstocks for fuels and chemicals from algae: Treatment of crude bio-oil over HZSM-5. *Algal Res.* **2013**, *2*, 154–163. [CrossRef]

26. Elliott, D.C.; Hart, T.R.; Schmidt, A.J.; Neuenschwander, G.G.; Rotness, L.J.; Olarte, M.V.; Zacher, A.H.; Albrecht, K.O.; Hallen, R.T.; Holladay, J.E. Process development for hydrothermal liquefaction of algae feedstocks in a continuous-flow reactor. *Algal Res.* **2013**, *2*, 445–454. [CrossRef]
27. Montgomery, D.C. *Design and Analysis of Experiments*, 8th ed.; John Wiley Sons: Singapore, 2012.
28. ASTM International. *ASTM D1160, Standard Test Method for Distillation of Petroleum Products at Reduced Pressure*; ASTM International: West Conshohocken, PA, USA, 2015.
29. ASTM International. *ASTM D2892, Standard Test Method for Distillation of Crude Petroleum (15- Theoretical Plate Column)*; ASTM International: West Conshohocken, PA, USA, 2005.
30. Maxwell, J.B.; Bonnell, L.S. Derivation and Precision of a New Vapor Pressure Correlation for Petroleum Hydrocarbons. *Ind. Eng. Chem.* **1957**, *49*, 1187–1196. [CrossRef]
31. Channiwala, S.A.; Parikh, P.P. A unified correlation for estimating HHV of solid, liquid and gaseous fuels. *Fuel* **2002**, *81*, 1051–1063. [CrossRef]
32. ASTM International. *ASTM D7169, Standard Test Method for Boiling Point Distribution of Samples with Residues Such as Crude Oils and Atmospheric and Vacuum Residues by High Temperature Gas Chromatography*; ASTM International: West Conshohocken, PA, USA, 2011.
33. Pedersen, T.H.; Jensen, C.U.; Sandström, L.; Rosendahl, L.A. Full characterization of compounds obtained from fractional distillation and upgrading of a HTL biocrude. *Appl. Energy* **2017**, *202*, 408–419. [CrossRef]

Article

Evaluating the Impacts of ACP Management on the Energy Performance of Hydrothermal Liquefaction via Nutrient Recovery

Sarah K. Bauer [1,†], Fangwei Cheng [2,†] and Lisa M. Colosi [2,*]

1 Department of Civil and Environmental Engineering, Rowan University, 201 Mullica Hill Road, Glassboro, NJ 08028, USA; bauers@rowan.edu

2 Department of Engineering Systems and Environment, University of Virginia, 351 McCormick Road, Charlottesville, VA 22904, USA; fc4uk@virginia.edu

* Correspondence: lmc6b@virginia.edu; Tel.: +1-434-924-7962; Fax: +1-434-982-2951

† These authors contributed equally to this paper.

Received: 15 January 2019; Accepted: 19 February 2019; Published: 22 February 2019

Abstract: Hydrothermal liquefaction (HTL) is of interest in producing liquid fuels from organic waste, but the process also creates appreciable quantities of aqueous co-product (ACP) containing high concentrations of regulated wastewater pollutants (e.g., organic carbon, nitrogen (N), and phosphorus (P)). Previous literature has not emphasized characterization, management, or possible valorization of ACP wastewaters. This study aims to evaluate one possible approach to ACP management via recovery of valuable scarce materials. Equilibrium modeling was performed to estimate theoretical yields of struvite ($MgNH_4PO_4 \cdot 6H_2O$) from ACP samples arising from HTL processing of selected waste feedstocks. Experimental analyses were conducted to evaluate the accuracy of theoretical yield estimates. Adjusted yields were then incorporated into a life-cycle energy modeling framework to compute energy return on investment (EROI) for the struvite precipitation process as part of the overall HTL life-cycle. Observed struvite yields and residual P concentrations were consistent with theoretical modeling results; however, residual N concentrations were lower than model estimates because of the volatilization of ammonia gas. EROI calculations reveal that struvite recovery is a net-energy producing process, but that this benefit offers little to no improvement in EROI performance for the overall HTL life-cycle. In contrast, corresponding economic analysis suggests that struvite precipitation may be economically appealing.

Keywords: hydrothermal liquefaction; aqueous co-product; nutrient recovery; struvite; EROI

1. Introduction

Hydrothermal liquefaction (HTL) is an appealing platform for the production of drop-in fuels from abundant wet, organic "waste" feedstocks [1–3]. However, HTL-derived biofuels are known to have a significantly high water footprint, arising from the large volumes of aqueous co-product (ACP) that are produced as secondary waste during the conversion process [4–6]. Until recent years, existing studies have placed relatively little emphasis on ACP quality and management. HTL studies that do address ACP quality primarily pertain to the liquefaction of various microalgae [7–11]. These studies have evaluated the feasibility of reusing ACP as a nutrient-rich growth medium for algae cultivation [7,8]. Few studies have characterized ACP quality arising from non-algae feedstocks [4]. All such studies, however, reveal that ACP contains very high concentrations of regulated constituents, including dissolved organic carbon (DOC), total nitrogen (TN), and total phosphorus (TP) [2,4,5,12]. Accordingly, widespread commercialization of HTL could create very large quantities of ACP requiring significant treatment and/or dilution prior to discharge into a receiving water body. The need for

ACP treatment could undermine the energetic favorability of the HTL platform, because wastewater treatment plant (WWTP) processes, most notably nutrient removal, are highly energy consuming [13]. Previous life-cycle modeling has demonstrated that energy consumption for ACP management could be on the same order of magnitude as energy consumption for liquefaction itself [4]. Thus, there is a need to harmonize water and energy sustainability objectives during HTL processing.

Separately, there is a strong need to increase agricultural production of food and fuel crops over the next few decades, which will likely exacerbate existing nutrient scarcity [14]. Thus, there is an increasing interest in recovering nitrogen (N) and phosphorus (P) from various wastewaters to offset dwindling availability of nutrients from virgin sources [15–17]. Many studies have demonstrated the feasibility of chemical precipitation as a means to recover nutrients from high-strength agricultural or industrial wastewaters. Several of these studies have focused on the recovery of struvite ($MgNH_4PO_4{\cdot}6H_2O$), a slow-release fertilizer, from concentrated wastewaters, including: landfill leachate, slaughterhouse wastewater, anaerobic digester effluent, swine and dairy manure wastewater, source-separated urine, and others [14,16,18–21]. These studies have illustrated the feasibility of struvite recovery from potent wastewaters and have laid the groundwork for understanding how reaction conditions (e.g., molar ratios, pH, competing reactions, etc.) impact process yield of struvite.

It was hypothesized that ACP arising from the HTL processing of various organic waste feedstocks may be suitable for nutrient recovery via struvite precipitation. This approach would be desirable for reducing the energy intensiveness of ACP management, while simultaneously producing valuable scarce materials in the form of slow-release fertilizer. Creation of a useful fertilizer by-product could also improve the economic favorability of the HTL platform. Accordingly, this study had three aims: (1) evaluate the technical feasibility of struvite precipitation from ACP samples arising from the HTL processing of select organic waste feedstocks, (2) evaluate whether struvite recovery compensates for the energy penalty of ACP management as part of the overall HTL life-cycle, and (3) assess the economic feasibility of struvite recovery from post-HTL ACP samples.

2. Materials and Methods

2.1. HTL Conversion and Characterization of ACP

Seven non-food, organic waste feedstocks were evaluated in this study: (1) pre-digested WWTP sludge; (2) digested WWTP sludge; (3) dairy manure; (4) brewery yeast and (5) spent grains from craft beer production; and (6) red lees and (7) white lees (i.e., yeast particles) from wine production. All samples were collected, thoroughly homogenized via blending, characterized, and stored at 4 °C prior to HTL processing. Raw waste feedstocks were characterized for total suspended solids (TSS), volatile solids (VS), and water and ash contents according to APHA Standard Methods (see Supplementary Materials, Table S10) [22]. Waste feedstocks were then hydrothermally liquefied in a 300-mL Parr Instrument Company (Moline, IL, USA) stirred reactor with a quartz liner and external heater, as previously described in Bauer et al. (2018) [4]. In short, 100 g of wet feedstock paste, adjusted to an optimal water content of 90% (m/m) to be consistent with previous literature [5], was pressurized to 100 psi with nitrogen (N_2) gas, continuously stirred at 300 rpm, and heated to 300 ± 5 °C at ~8–10 °C/min. Once heated, the reactor was maintained at 300 ± 5 °C for a residence time of 30 min [4,7,8]. All characterization analyses and HTL reactions were conducted in triplicate.

HTL product phases (i.e., biofuel, biochar, and ACP) were separated using techniques adapted from Xu and Savage (2014) [23], as previously described in Bauer et al. (2018) [4]. In short, solid-phase biochar was separated from the liquid phase via filtration. Biofuel and ACP were separated from each other via extraction into dichloromethane (DCM) at a ratio of 1–2x vol HTL liquid/vol DCM. Reactor contents were then decanted and centrifuged to facilitate phase separation, and the ACP was manually drawn off. Liquid biocrude was operationally defined based on solubility in DCM. The solvent was then evaporated using a gentle stream of N_2 gas over a 24-h period. Resulting ACP was filtered using a 0.22-μm pore-size filter. Post-HTL ACP was characterized using APHA

Standard Methods or commercial HACH kits based on traditional wastewater parameters, including pH, TN, TP, ammonium (NH_4^+), and orthophosphate (PO_4^{3-}) methods (see Supplementary Materials, Section 5.0) [22]. ACP quality was further characterized using a Thermo Fisher Scientific Dionex ICS 5000 DP-5 Ion Chromatograph (IC) (Waltham, MA, USA), using U.S. EPA Method 300.1 [24], with a detection limit of 0.5 mg/L for: Mg^{2+}, Ca^{2+}, K^+, Na^+, Cl^-, NO_3^-, NO_2^-, and SO_4^{2-}. Additionally, titration experiments with 1 N sodium hydroxide (NaOH) solution were performed to measure the quantity of NaOH required to increase the pH of each post-HTL ACP to a required value.

2.2. Theoretical Recovery of Nutrients from ACP

A chemical equilibrium modeling software package, U.S. EPA Visual MINTEQ Version 3.1 (Stockholm, Sweden) [25], was used to estimate the theoretical recoverability of N and P from the various ACP samples arising from the HTL processing of selected organic waste feedstocks. Measured pH and constituent concentrations (e.g., NH_4^+, PO_4^{3-}, and various anions and cations referenced in Section 2.1) were entered into the software individually for each post-HTL ACP in order to assess possible competing reactions to struvite precipitation. Equilibrium modeling was then applied in two ways. First, pH was artificially increased from its initial measured value, in intervals of 0.5 pH units, up to a maximum value of 14. Second, magnesium (Mg^{2+}) concentration was artificially increased from its initial measured value, in intervals of 5 mg/L, for 500 intervals. For each interval step of both virtual titrations, the model computed what masses of struvite and other relevant solids would precipitate and then estimated the residual dissolved-phase concentrations of the various constituents, including relevant complexes. Optimal Mg^{2+} doses for each ACP sample were selected based on consumption of Mg^{2+} to produce the desired product. That is, Mg^{2+} was added stepwise up to the point where $\geq$50% of the last dose remained in solution or was diverted to a competing side reaction that did not produce struvite. Corresponding pH values were selected as optimal for each ACP.

2.3. Experimental Recovery of Nutrients from ACP

Modeling estimates for struvite yield and residual TN and TP concentrations were evaluated experimentally using protocols from Ishii and Boyer (2015), Yetilmezsoy and Sapci-Zengin (2009) [16,20]. Reactors comprised duplicate 50-mL beakers containing 20 mL of ACP. pH was adjusted via dropwise addition of 1 N NaOH up to optimal pH values identified via modeling results. Magnesium chloride ($MgCl_2{\cdot}6H_2O$) was used as a Mg^{2+} source at a Mg^{2+}:PO_4^{3-} molar ratio of 1.5:1 to facilitate the precipitation of struvite [26]. Reactors were stirred at 200 rpm for 30 minutes at 20 °C, settled quiescently for 30 minutes, and then filtered using a 0.22-µm pore-size filter. Resulting solids were dried at 20 °C and weighed daily until the mass of the solids remained unchanged (~3 days). Residual ACP concentrations of NH_4^+, PO_4^{3-}, and Mg^{2+} were measured using commercial HACH kits. Solid precipitates were analyzed by two methods. All precipitates were dissolved individually in concentrated hydrochloric acid (HCl), and the concentrations of NH_4^+, PO_4^{3-}, and Mg^{2+} were analyzed using commercial HACH Company (Loveland, CO, USA) water quality analysis kits. A portion of precipitates were further analyzed by X-ray diffraction (XRD) using a Panalytical Empyrean Diffractometer equipped with a Bragg-Bretano HD Prefix module and a GaliPix3D Area Detector operating in scanning line mode. This analysis was used to confirm the presence of struvite in the solid precipitates from post-HTL ACPs. Crystalline solid precipitates were scanned for 2-theta = 5 to 70° at a rate of 4 rps and repeated four times in order to improve the signal to noise ratio.

2.4. Energy Analysis of ACP Management via Struvite Precipitation

Life-cycle analysis (LCA) was used to assess the energy return on investment (EROI), which expresses the relationship between energy produced (E_{OUT}) and energy consumed (E_{IN}) (i.e., EROI = E_{OUT}/E_{IN}) for ACP management via nutrient recovery in the form of struvite precipitation. These calculations accounted for the inputs and outputs of the precipitation process, including: materials consumption for chemical precipitation, production of the fertilizer product, and reduced need for energy-intensive WWTP removal to remediate ACP. Energy consumption for post-precipitation solids handing (e.g., pumping, drying, and/or purification) was not evaluated. Embedded energy contents for precipitation materials were collected from the ecoinvent database, as accessed using SimaPro v.3 (Amersfoort, Netherlands), and/or adapted from Clarens et al. (2010) [13]. Energy input (i.e., E_{IN}) estimates accounted for the energy costs of NaOH (46.6 kJ/g) and $MgCl_2$ (4.7 kJ/g) based on titration to achieve model-designed optimal pH and an assumed Mg^{2+}:$PO_4{}^{3-}$ molar ratio of 1.5:1 (see Supplementary Materials, Table S8).

Energy output (i.e., E_{OUT}) estimates accounted for two kinds of benefits: (1) the energy value of struvite precipitates, based on a presumption that it would displace some other commercial fertilizer with known embedded energy footprint; and (2) the energy offset arising from reduced need to apply energy-intensive WWTP treatments for the removal of TN and TP down to acceptable levels. Both calculations make use of a systems-expansion approach, whereby the value of a product in one system is computed by assuming that there is some offset burden in a separate, related system. The validity of LCA estimates computed using this approach are necessarily dependent on the validity of the assumed interactions between the systems. For this reason, a range of E_{OUT} values were computed for three different scenarios, as described below:

- Scenario 1: Energy output corresponds to only the amount of commercial fertilizer (i.e., monoammonium phosphate (MAP)) supplanted by struvite precipitation.
- Scenario 2: Energy output corresponds to MAP avoidance energy plus avoided energy costs for WWTP removal of TN and TP in ACP, assuming ammonia (NH_3) is not allowed to volatilize during struvite recovery (i.e., "closed system").
- Scenario 3: Energy output corresponds to MAP avoidance energy plus avoided energy costs for WWTP removal of TN and TP, assuming NH_3 is allowed to volatilize during struvite recovery (i.e., "open system").

For each scenario, it was assumed that recovered struvite could be used to replace the commercial fertilizer MAP on a 1:1 molar basis for both N and P, because both fertilizers have the same 1:1 N:P molar ratio. The values of 38 MJ/kg P_2O_5 and 58.7 MJ/kg N were used as the energy values of MAP, based on the ecoinvent database. The overall value following adjustment of units was 13.5 kJ/g MAP. The energy value of the struvite product was determined by multiplying the molar value of struvite yield by the molar mass of MAP (115.03 g/mol) and the energy value of MAP.

It was assumed that residual TN and TP in the post-HTL ACP would be removed via conventional WWTP methods. It was assumed that TN removal occurs via biological nitrification and separate-stage denitrification using methanol as a co-substrate (3.4 g methanol/g N removed) [27]. The energy intensity of methanol (38 kJ/g methanol) was obtained from Clarens et al. (2010) [13]. Using these values, the energy offset for avoided TN removal was 129.2 kJ/g N. It was assumed that TP removal occurs via chemical precipitation using ferrous sulfate ($FeSO_4$) [27]. The energy demand and energy consumption for $FeSO_4$ were obtained from Clarens et al. (2010) [13]: 1.8 g $FeSO_4$/g P removed and 1.95 kJ/g P removed, respectively. Using these values, the energy offset for avoided TP removal was 3.51 kJ/g.

Energy ratio metrics for the overall HTL life-cycle were collected from relevant literature [28–30]. None of the original studies accounted for ACP management and/or recovery of useful materials. Accordingly, the model frameworks for each of the original studies were modified to include experimental data from the present study, taking a similar approach as Bauer et al. (2018) [4]. The energy ratio metrics were then recomputed, as a means to assess to what extent ACP management via struvite recovery could affect the overall energy performance of the HTL platform.

2.5. Economic Analysis of ACP Management via Struvite Precipitation

A preliminary economic analysis was conducted in order to evaluate the economic favorability of struvite precipitation from post-HTL ACP samples. Calculations were scaled to the treatment of

1 million liters of post-HTL ACP. It had been previously demonstrated that chemical costs are the major driver for overall cost of the struvite recovery process [31]. As such, only the costs of raw chemical were considered in this study. Unit prices of $MgCl_2$, NaOH, methanol, and $FeSO_4$ were assumed to be \$240/ton, \$425/ton, \$500/ton, and \$120/ton, respectively. Struvite yields and residual TN and TP concentrations were based on experimental measurements from this study. Net income from the struvite precipitation process was evaluated for the same three scenarios outlined previously: (1) supplanted MAP only, (2) supplanted MAP + WWTP offsets (assuming a closed system), and (3) supplanted MAP + WWTP (assuming an open system).

3. Results and Discussion

3.1. Characterization of Post-HTL ACP

The goal of this study was to evaluate the recoverability of valuable nutrients (i.e., N and P) from ACP as a means of managing secondary waste created during HTL processing and potentially increasing the overall EROI of the HTL platform. Table 1 presents post-HTL ACP characterization data for the seven organic waste feedstocks included in this study. Several noteworthy observations from this data pertain to the initial pH values and concentrations of relevant constituents. Regarding pH, it is fortuitous that most feedstocks produce ACPs that are slightly basic: $\geq$8 for pre-digested sludge, digested sludge, brewing yeast, and red lees ACPs. Dairy manure and spent grains ACPs are moderately acidic. Relatively high starting pH values are favorable for the proposed nutrient recovery application, because less base (e.g., NaOH) is required to adjust the pH to the high range required for struvite precipitation. Regarding select constituents, all ACP samples contain appreciable to very high concentrations of TN and TP.

Table 1 also reveals that there is significant variability in the fractions of the TN and TP pools available for struvite precipitation. The fraction of TN present as NH_4^+, which is the form required for struvite precipitation, varies from 30–96%. Digested sludge and brewing yeast ACP samples exhibited the highest fractions of recoverable TN. Correspondingly, the fraction of TP present as PO_4^{3-}, which is the form required for struvite precipitation, varies from 8–100%. White lees ACP exhibited the highest measured fraction of TP as PO_4^{3-} (100%), whereas brewing yeast and spent grains ACPs exhibited the next highest fractions (33–34%). All other ACP samples exhibited very low fractions of TP as phosphate. These results are somewhat consistent with previous research. For example, Valdez et al. (2012) [9] investigated the hydrothermal processing of *Nannochloropsis sp.* under various reaction conditions. The average observed NH_4^+/TN ratio was 48%, which is the same as the average value from Table 1. On the other hand, this study observed that P was mostly present as OP, which was not the case for this study. An interesting conclusion of the Valdez et al. (2012) study was that biomass characteristics may be more important than HTL reaction conditions for NH_4^+/TN and OP/TP ratios [9]. This observation is important for understanding the generalizability of the results from the current study.

Finally, it is of interest to examine the ratios of recoverable N and P, to assess which reactant will limit struvite yield. All seven of the ACP samples evaluated are phosphate-limited. The white lees ACP exhibited a ratio of 1.3:1 NH_4-N:PO_4-P, but all other ACPs contained dramatic quantities of excess ammonium. This is not surprising considering the large ratio of N to P in most plant and microbial biomasses.

Table 1. Characterization of post-hydrothermal liquefaction (HTL) aqueous co-product (ACP) of select waste feedstocks for pH and various wastewater constituents. Average values are reported and used as inputs for modeling and experimental analysis. ND denotes measurements below the detection limit of 0.5 mg/L. All samples exhibited ND nitrite (NO_2^-).

Waste Feedstock	PH	TN (mM as N)	NH_4^+ (mM as N)	NH_4^-N/TN-N Ratio	TP (mM as P)	OP (mM as P)	OP-P/TP-P Ratio	NH_4^-N/OP-P Ratio	MG^{2+} (mM)	CA^{2+} (mM)	K^+ (mM)	NA^+ (mM)	CL^- (mM)	NO_3^- (mM)	SO_4^{2-} (mM)
Dairy Manure	4.4	75	23	0.30	15.4	1.3	0.08	17.9	2.2	7.5	7.7	6.1	8.8	ND	0.3
Pre-Digested Sludge	8.4	232	70	0.30	25.8	3.6	0.14	19.3	0.1	0.5	6.9	3.0	2.1	ND	2.4
Digested Sludge	8.6	156	150	0.96	7.1	1.0	0.14	150.0	0.2	0.8	7.6	4.3	2.8	ND	3.6
Brewing Yeast	8.3	175	98	0.56	70.8	24.3	0.34	4.0	0.2	0.5	33.9	1.7	2.7	ND	3.1
Spent Grains	5.3	146	50	0.34	33.5	11.4	0.34	4.4	1.2	0.4	0.3	1.3	0.1	0.04	0.6
White Lees	6.4	7	2	0.32	1.5	1.7	1.18	1.3	0.2	0.2	81.2	0.4	0.2	0.02	0.4
Red Lees	8.8	135	80	0.59	117.2	22.2	0.19	3.6	0.1	0.5	182.0	0.7	0.3	ND	ND

Struvite is produced when Mg^{2+}, NH_4^+, and PO_4^{3-} combine together at a molar ratio of 1Mg:1N:1P under high pH. However, these constituents can also interact with other ions (e.g., Ca^{2+}, K^+, Na^+, Cl^-, SO_4^{2-}, etc.) in competing side reactions. The relative favorability of struvite precipitation compared to other possible reactions is a function of pH and various constituent concentrations [32]. Table 1 summarizes concentrations of relevant ions present in the post-HTL ACP samples. Among all of the evaluated ACPs, nitrate (NO_3^-) and nitrite (NO_2^-) (not shown in Table 1) are present in negligible amounts. In contrast, most ACP samples contain relatively high concentrations of K^+, Na^+, Cl^-, and SO_4^{2-}. Only dairy manure and spent grains exhibit appreciable Mg^{2+} concentrations, which means that most samples will require full dosing with Mg^{2+} to completely remove either NH_4^+ or PO_4^{3-} (whichever is limiting). Finally, Ca^{2+} is known to form several precipitates with PO_4^{3-} and/or other constituents at high pH (e.g., $Ca(H_2PO_4)_2$, $CaHPO_4$, etc.). Formation of these solids reduces phosphate availability, which decreases the theoretically achievable struvite yield given that phosphate is the limiting reactant for most feedstock ACPs evaluated in this study. Although the various calcium phosphate salts are useful as agricultural fertilizers, these are less desirable than struvite within the specific context of ACP management, because it is of interest to remove as much TN as possible while also removing TP, especially since TP is the limiting reactant. Removing TP without also removing TN leaves greater residual TN concentrations, such that more additional post-treatment will be required for the ACP. Given all of the various competing reactions that can occur, it is of interest to see if/how the various combinations of constituents in ACP samples can affect precipitation and residual concentrations of TN and TP.

3.2. Theoretical Recovery of Nitrogen and Phosphorus from ACP

Equilibrium pH modeling was used to determine the theoretical recovery of dissolved nutrients from the various post-HTL ACP samples, with a goal of maximizing struvite precipitation. The model was parameterized separately for each ACP sample, based on data from Table 1. Modeling results are summarized in Table 2.

Based on modeling results for each ACP sample, Table 2 presents the optimum pH values for struvite recovery and the corresponding quantity of NaOH required to achieve that value; the theoretical maximum amount of struvite precipitate that can be produced; and the theoretical estimates of residual NH_4^+/NH_3 and orthophosphate (OP) concentrations after struvite has been recovered. The data in Table 2 reveal several important findings. First, pertaining to nutrient recovery, the model predicts that struvite recovery is theoretically possible for all but one feedstock ACP. The only exception is the dairy manure ACP, for which the high concentration of Ca^{2+} in the ACP results instead in considerable precipitation of HAP (i.e., hydroxyapatite ($Ca_5(PO_4)_3(OH)$)) at high pH. The creation of HAP as a competing solid depletes the available phosphate and prevents the formation of struvite. Accordingly, TN removal is 0% even though orthophosphate removal is approximately 100%. Second, phosphate is almost completely consumed for all ACP samples, leaving very low residual concentrations. This outcome is consistent with previous efforts to recover struvite from various potent wastewaters [14,21]. This result is also consistent with the observation that all ACP samples are phosphate-limited, as evident in Table 1. The only ACP for which phosphate removal is not essentially 100% is the white lees ACP. As shown in Table 2, the model predicts significant residual concentrations for both NH_4^+/NH_3 and OP for the white lees ACP. This outcome is a pH effect related to the speciation of NH_4^+/NH_3 versus $PO_4^{3-}/HPO_4^{2-}/H_2PO_4^-/H_3PO_4$ (i.e., the various forms of orthophosphates). As noted above, struvite formation requires a combination of NH_4^+ and PO_4^{3-}, but there is a very small pH window over which these two species exist together. The pK_A for NH_4^+/NH_3 is 9.25, and the pK_A for HPO_4^2/PO_4^{3-} is 12.35 [33]. Thus, increasing the pH above 9.25, which is useful for increasing struvite yields in all other ACP samples because they are so strongly OP-limited, reduces NH_4^+ availability by converting it to NH_3 (see Supplementary Materials, Figure S1; Tables S1–S7). Therefore, the specific N:P ratio exhibited by the white lees ACP, which is much lower than the ratios exhibited by all other ACP samples, make it such that neither NH_4^+/NH_3 nor OP can be completely removed via struvite precipitation.

Table 2. Theoretical estimates of struvite recovery and residual total nitrogen (TN) and total phosphorus (TP) concentrations from post-HTL ACPs of select waste feedstocks based on equilibrium chemistry modeling using Visual MINTEQ.

Waste Feedstock	Optimal pH	NaOH Consumed (mM)	$MgCl_2$ Consumed (mM)	Struvite Recovered (mM)	OP Removed (mM)	NH_4^+/NH_3 Removed (mM)	Residual NH_4^+/NH_3 (%)	Residual TN (%)	Residual OP (%)	Residual TP (%)
Dairy Manure	8.0	1.40	0	0	1.27	0	100	100	0	75
Pre-Digested Sludge	10.5	1.00	3.29	3.28	3.59	3.28	95	98	>1	57
Digested Sludge	10.5	1.50	0.41	0.48	0.97	0.48	100	100	2	58
Brewing Yeast	10.5	1.14	23.9	23.9	24.2	23.9	76	82	>1	5
Spent Grains	10.5	1.35	9.88	11.1	11.3	11.1	78	90	1	3
White Lees	9.0	0.09	3.70	0.56	0.69	0.56	75	90	34	36
Red Lees	10.5	1.12	21.8	21.8	22.1	21.8	73	79	>1	42

3.3. Experimental Recovery of Nitrogen and Phosphorus from ACP

Precipitation experiments were performed to validate theoretical modeling results. Figure 1 summarizes a comparison between experimental results and theoretical predictions, including struvite mass yields and removal efficiencies for OP and NH_4^+. As expected, experimental analysis confirmed that struvite is not produced from the dairy manure ACP; therefore, the dairy manure ACP is excluded from Figure 1 and subsequent results. For all other feedstock ACPs, experimentally measured struvite mass yields were highly consistent with theoretical predictions from the model, as seen in Figure 1a. XRD analysis confirmed the chemical identity of crystalline struvite for the spent grains, brewing yeast, and red lees ACPs. The crystalline form of struvite agrees with previously published literature [21,34]. All precipitates were also dissolved in strong acid and analyzed for NH_4^+, PO_4^{3-}, and Mg^{2+} using commercial HACH kits. The ratio of these measured constituents further confirmed that these solids were predominantly struvite, with relatively low concentrations of impurities (see Supplementary Materials, Figure S3; Table S9). Across all the ACP samples, the average molar ratio for Mg^{2+}:NH_4^+:PO_4^{3-} was 1.1:1:1.2. This ratio is consistent with previously published studies of nutrient recovery via struvite precipitation with reasonably low impurity content [34]. Overall, the good correspondence between modeling results and experimental struvite yields confirms the usefulness of the modeling approach for estimating how much struvite can be produced from various ACP samples with known constituent concentrations.

Experimental results also confirmed the accuracy of theoretical estimation of OP removal via struvite precipitation. This is not surprising given the good accuracy of the predicted struvite mass yields, since all of the ACP samples were phosphate-limited (Figure 1b). However, the theoretical predictions of NH_4^+ removal and residual concentration were much less accurate than for OP. As seen in Figure 1c, the modeling predictions dramatically underestimated NH_4^+ removal for all evaluated ACP samples. This unexpected removal is attributed to NH_3 volatilization during the period of time when struvite particles were quiescently settling in an open reactor at room temperature. From literature, it has been previously demonstrated that up to 40% of residual NH_4^+/NH_3 may be volatilized following pH adjustment to promote struvite precipitation from various potent wastewaters. For example, Çelen et al. (2007) reported average NH_4^+ losses of 35–40% during experimental precipitation of struvite at a pH of 8.5 [21]. Rahman et al. (2011) documented a range of NH_4^+ losses of 26.5–29.4% via NH_3 volatilization [19]. In this study, the magnitude of NH_4^+ removal occurring via volatilization was as large as or much larger than the magnitude of removal occurring via struvite precipitation.

Table 3 summarizes pre- and post-precipitation of N and P concentrations in dissolved phase and as struvite precipitates. These data are useful for assessing the mass balances for PO_4^{3-}, NH_4^+, and Mg^{2+} during precipitation reactions. As seen in Table 3, the mass balances for PO_4^{3-} close to within 5% on average across all ACP samples tested. The mass balances close similarly well for Mg^{2+}, with greater than 90% of the original dosing accounted for among the solid and dissolved phase products. However, the NH_4^+ mass balances for the ACP samples do not close well, with nearly 30% NH_4^+ on average unaccounted for in the solid and dissolved phase products. As mentioned above, the missing NH_4^+ is thought to have volatilized as gaseous NH_3 during quiescent settling.

Table 3. Experimental results for pre- and post-precipitation concentrations of nitrogen (N) and phosphorus (P), with additional mass balance information for PO_4^{3-}, NH_4^+, and magnesium (Mg^{2+}).

Waste Feedstock	Initial Concentration (mg/L)			Recovered as Precipitates (mg/L)			Remaining in ACP (mg/L)			Mass Difference (%)		
	PO_4^{3-}-P	NH_4^+-N	Mg^{2+}	PO_4^{3-}-P	NH_4^+-N	Mg^{2+}	PO_4^{3-}-P	NH_4^+-N	Mg^{2+}	PO_4^{3-}-P	NH_4^+-N	Mg^{2+}
Pre-Digested Sludge	112	975	121	106	38	73	5.6	620	40	0.8	32.5	6.4
Digested Sludge	31	2100	36	29	13	25	0.0	1528	11	5.2	26.6	1.0
Brewing Yeast	753	1370	875	751	285	503	1.6	568	285	0.0	37.8	9.9
Spent Grains	352	700	409	324	122	218	0.0	422	190	8.1	22.3	0.2
White Lees	53	31	62	30	13	13	18.0	10	31	10.3	27.4	11.5
Red Lees	687	1115	798	646	266	521	0.0	587	204	6.0	23.6	9.1
Average	331	1048	385	314	123	227	4.2	622	127	5.1	28.4	8.0

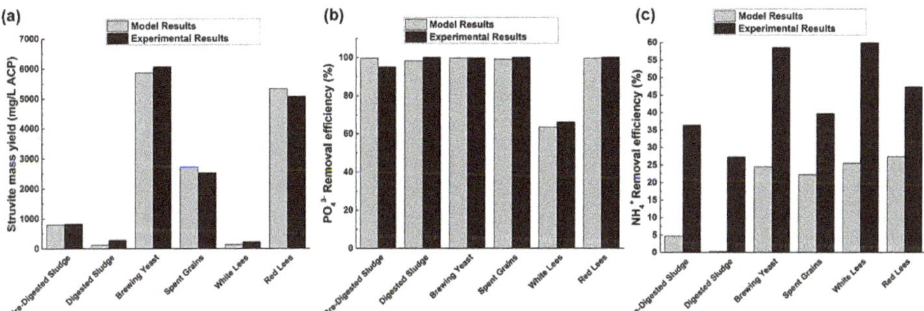

Figure 1. Theoretical and experimental mass yield of struvite (**a**) and removal efficiency of (**b**) orthophosphate (PO_4^{3-}) and (**c**) ammonium (NH_4^+) from post-HTL ACP of select waste feedstocks via precipitation of struvite.

3.4. Energy Recovery Impacts of HTL Processing

It is also of interest to quantify the energetic favorability of the struvite precipitation process. This was done by computing EROI (i.e., E_{OUT}/E_{IN}) metrics, assuming that the ACP arising from HTL processing of each waste feedstock is managed via struvite precipitation followed by conventional treatment in a municipal WWTP. Figure 2 summarizes the results of this analysis for three scenarios (as defined in Section 2.4). For Scenario 1, which is the most conservative scenario, EROI values were found to be ranging from 0.4 to 1.0. This indicates that struvite precipitation from ACP is not energetically favorable when the energy value of the fertilizer (i.e., struvite) product is considered to be the only energy product. This result occurs because the combined energy costs for the Mg salt (i.e., $MgCl_2$) and NaOH (46.6 kJ/g and 4.7 kJ/g, respectively) required to induce struvite precipitation are comparable to the energy value of the fertilizer product itself (13.5 kJ/g assuming direct replacement for struvite). For Scenario 2, EROI values increase to 0.9–2.1. These estimates account for both the energy value of the struvite as fertilizer replacement and also the avoided WWTP burden arising from not having to remediate N and P that is transformed into struvite solids. For both Scenarios 1 and 2, there is strong correlation between the amount of struvite recovered and the EROI value, with increases in struvite mass yield corresponding to increases in EROI. Finally, for Scenario 3, EROI values increase dramatically to 2.1–14.1. This increase accounts for the high energy intensity associated with nutrient removal during conventional wastewater treatment. In this scenario, the energy consumption for avoided TN removal becomes the determining factor for EROI. Since residual TN removal is energy- and cost-intensive, and the volatilization of NH_3 reduces the need for the removal of TN post-struvite precipitation in a WWTP, EROI values for this scenario are much higher than in Scenarios 1 and 2.

Previously published EROI values for HTL processing were revised to account for ACP management based on the experimental results of this study. Original HTL values were taken from Connelly et al. (2015), Sawayama et al. (1999), and Vardon et al. (2012) [28–30]. These original EROI estimates were previously revised by Bauer et al. (2018) to account for ACP management via conventional wastewater treatment for removal of TN, TP, and COD [4]. In the Bauer et al. (2018) study, the original EROI values were revised to account for the energy consumption required to remove TN, TP, and BOD from the post-HTL ACPs through conventional wastewater treatment processes. In the current study, revised EROI estimates from Bauer et al. (2018) were re-revised to account for nutrient recovery from the ACPs via struvite precipitation, assuming that residual (i.e., post-precipitation) TN, TP, and BOD in the ACPs are removed via conventional treatment in a municipal WWTP [4].

Table 4 summarizes the original EROI values from the three studies, the revised EROI values presented in Bauer et al. (2018) [4], and re-revised EROI values that have been updated to account for N and P removal via struvite precipitation. Figure S2 in the Supplementary Materials illustrates the life-cycle systems boundaries for the three sets of EROI calculations. Neither Bauer et al. (2018) [4] nor this study attempted to change or harmonize the systems boundaries of the original studies beyond adding in ACP management with or without nutrient recovery. The original analyses were changed as little as possible. Therefore, the EROI values in Table 4 should be compared only within a single row (e.g., comparing one original EROI with its corresponding revised or re-revised EROI), rather than comparing EROIs from multiple rows. The latter is not appropriate given that the original calculations made use of different systems boundaries.

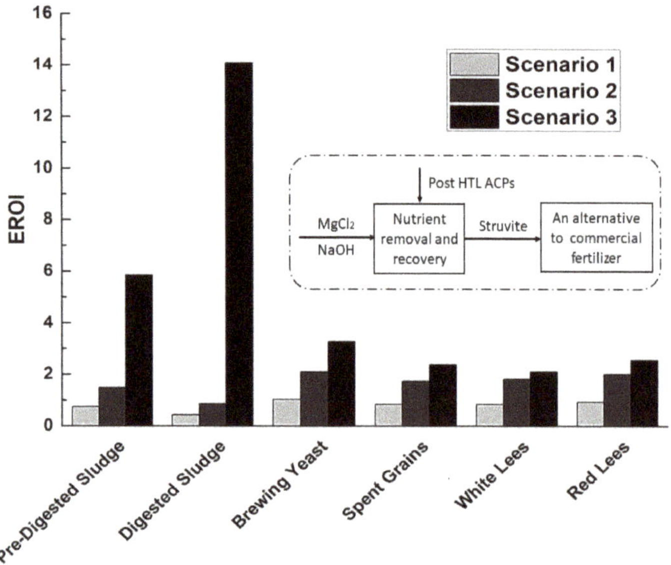

Figure 2. Calculated energy return on investment (EROI) values of nutrient removal and recovery of post-HTL ACPs of select waste feedstocks via struvite recovery.

Table 4. Three sets of EROI values for HTL processing of various organic feedstocks: as reported in original HTL studies; as revised by Bauer et al. (2018) [4] to include ACP management in a conventional wastewater treatment plant (WWTP); and as re-revised in this study to account for nutrient recovery via struvite precipitation with subsequent conventional treatment to remove residual TN, TP, and biological oxygen demand (BOD).

Author/Year	Feedstock/Scenario	Original EROI	Revised EROI with ACP Management [a]	Revised EROI with Struvite Precipitation [b]
Connelly et al., 2015 [28]	Algae, "CO_2 from ethanol"	1.3	1.1	1.2
	Algae, "CO_2 from natural wells"	1.2	1.0	N/A [c]
Sawayama et al., 1999 [30]	*B. braunii* (algae)	6.7	3.7	3.9
	D. tertiolecta (algae)	2.9	2.0	2.1
	Japanese oak	1.8	1.3	1.3
	Japanese larch bark	0.9	0.8	0.8
	Sewage sludge	2.9	2.0	2.0
	Barley silage	2.3	1.7	1.7
	Kitchen garbage	0.7	0.6	0.6
Vardon et al., 2012 [29]	*Scenedesmus* (algae), 80% moisture	2.3	1.3	1.4
	Defatted *Scenedesmus*, 80% moisture	1.8	1.1	1.1
	Spirulina (algae), 80% moisture	1.6	0.9	1.0

[a] Bauer et al. (2018) [4]; [b] This Study; [c] N/A due to non-dimensional scaling.

The calculations from this study reveal that the re-revised EROI values are not appreciably greater than the revised EROI values from Bauer et al. (2018) [4]. That is, accounting for struvite precipitation does not dramatically change the energy balance for the HTL system when accounting for ACP management, even when the best case scenario is applied. This indicates that the precipitation of struvite is not a highly promising way of improving the life-cycle energy performance of a waste-to-energy HTL platform. Several reasons for the poorer than expected performance of struvite precipitation include: the small fractions of ACP N and P that exist in the forms needed to produce struvite (i.e., NH_4^+ and PO_4^{3-}, respectively); and the low energy value of the MAP fertilizer which is supplanted by the struvite product, relative to the energy cost of the $MgCl_2$ that is used to produce it.

3.5. Economic Benefit of Struvite Precipitation from ACP

Based on the experimental results from this study, an economic analysis was conducted to measure the economic benefit of struvite precipitation from post-HTL ACP as a means of ACP management based on the three scenarios described in Section 2.4. Again, calculations were scaled to treatment of 1 million liters of post-HTL ACP. Based on measured ratios of biocrude:ACP, this volume of ACP would correspond to the production of approximately 110 thousand L liters of biocrude (see Supplementary Materials, Table S11). The market value of struvite has been previously reported by several researchers to be within the range of $198/ton to $800/ton [35]. For this analysis, the lowest reported market value for struvite ($198/ton) was used. The results from the economic benefit analysis of the precipitation of struvite from post-HTL ACP samples are summarized in Figure 3.

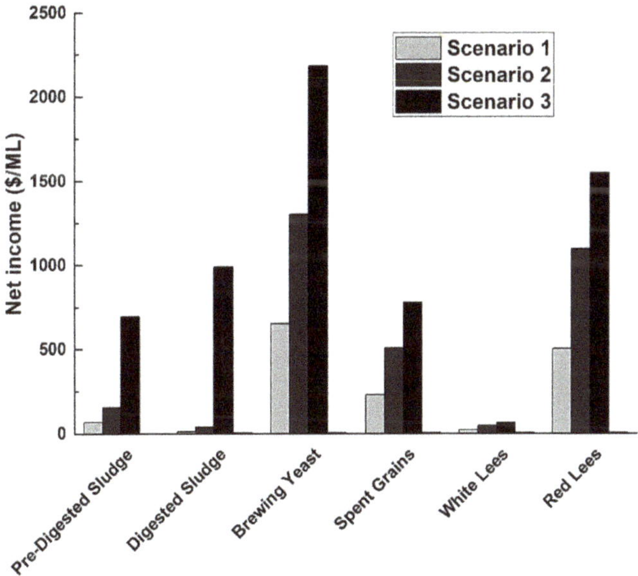

Figure 3. Economic analysis of the benefit of the precipitation of struvite from post-HTL ACP of select waste feedstocks as a means of management of ACP through the removal of NH_4^+ and PO_4^{3-}.

Based on the results reported in Figure 3, the economic benefit of struvite precipitation from post-HTL ACP is promising. All feedstock ACPs showed a positive net income from the recovery of N and P via the precipitation of struvite, ranging from $9/million liters (ML) ACP to $652/ML (Scenario 1), from $38/ML to $1298/ML (Scenario 2), and from $64/ML to $2179/ML (Scenario 3), even though the lowest reported market value of struvite was used for this analysis. The net income of struvite production is positively correlated to the concentration of PO_4^{3-} in the post-HTL ACP that can be recovered via nutrient-based precipitation. Compared to the concentration of recoverable (i.e., organic) N and P in benchmark domestic and industrial wastewaters, the concentrations of recoverable N and P in post-HTL ACP are much higher [4], indicating that the recovery of nutrients from post-HTL ACPs via struvite precipitation would be more economical than conventional wastewater treatment methods.

4. Conclusions

HTL processing of organic waste feedstocks produces potent wastewater in the form of ACP that is likely to require management before discharge into receiving water bodies. Theoretical modeling results indicate that inorganic N and P can be recovered from post-HTL ACP in the form of the slow-release fertilizer struvite. Most ACPs exhibited theoretical precipitation of struvite at an optimal pH of 10.5. Struvite precipitation accounted for more than 97% OP recovery for several of the studied ACP samples. Experimental analyses confirmed predicted struvite mass yields and estimated residual phosphate concentrations; however, NH_4^+ was even greater than expected, due to NH_3 volatilization. Although the struvite precipitation step is itself net-energy producing, adjustment of published EROI values for the HTL platform reveals that struvite precipitation does not appreciably change the overall energy balance of HTL. Economic analysis suggests that the precipitation of struvite from post-HTL ACP is economically feasible. This research provides essential insight into the sustainability of waste-to-energy systems as a means of both producing renewable energy and recovering valuable materials.

Supplementary Materials: The following are available online at http://www.mdpi.com/1996-1073/12/4/729/s1, Table S1: Theoretical recovery of nutrients via the precipitation of struvite from post-HTL dairy manure ACP at pH range of 4.4 to 14, Table S2: Theoretical recovery of nutrients via the precipitation of struvite from post-HTL pre-digested sludge ACP at pH range of 8.4 to 14. Optimal pH for nutrient recovery from pre-digested sludge ACP is 10.5, Table S3: Theoretical recovery of nutrients via the precipitation of struvite from post-HTL digested sludge ACP at pH range of 8.6 to 14. Optimal pH for nutrient recovery from digested sludge ACP is 10.5, Table S4: Theoretical recovery of nutrients via the precipitation of struvite from post-HTL brewing yeast ACP at pH range of 8.3 to 14. Optimal pH for nutrient recovery from brewing yeast ACP is 10.5, Table S5: Theoretical recovery of nutrients via the precipitation of solids from post-HTL spent grains ACP at pH range of 5.3 to 14. Optimal pH for nutrient recovery from spent grains ACP is 10.5, Table S6: Theoretical recovery of nutrients via the precipitation of solids from post-HTL white lees ACP at pH range of 6.4 to 14. Optimal pH for nutrient recovery from white lees ACP is 9.0, Table S7: Theoretical recovery of nutrients via the precipitation of solids from post-HTL red lees ACP at pH range of 8.8 to 14. Optimal pH for nutrient recovery from red lees ACP is 10.5, Table S8: Summary of the life-cycle energy values of the various materials consumed and produced via the precipitation of nutrients from the post-HTL ACP of select organic waste feedstocks. Parameter values were taken from the ecoinvent database, as accessed using SimaPro v.3 and/or Clarens et al. (2010), Table S9: Molar ratio of P-PO_4^{3-}, N-NH_4^+, and Mg^{2+} within the solids precipitated from the post-HTL ACP of select waste feedstocks, Table S10: Characterization of as-received raw waste feedstocks for processing into liquid biocrude through HTL conversion, as adapted by Bauer et al. (2018), Table S11: Biocrude and ACP yield from HTL process for select waste feedstocks, as adapted by Bauer et al. (2018), Figure S1: pH dependence of (a) OP and (b) ammonia, magnesium, and calcium speciation at 25 °C. The speciation of ammonia between ionized NH_4^+ and free NH_3, as well as the OP species (i.e., H_3PO_4, $H_2PO_4^-$, HPO_4^{2-}, and PO_4^{3-}) is pH dependent, Figure S2: Life cycle boundaries for three estimates of HTL EROI are denoted A, B and C. "A" (dashed black lines) depicts initial systems boundaries from original HTL studies (i.e., Connelly et al. (2015), Vardon et al. (2012) and Sawayama et al. (1999)). "B" (dashed green lines) depicts revised system boundaries used by Bauer et al. (2018) to account for ACP treatment in a municipal WWTP. "C" (dashed maroon lines) depicts extended system boundaries used in this study to account for struvite recovery followed by conventional treatment to remove residual TN, TP and BOD, Figure S3: XRD diffractogram of standard struvite and solids precipitated from the post-HTL red lees, brewing yeast and spent grains ACP samples.

Author Contributions: The individual contributions of the authors are as follows: conceptualization, S.K.B., F.C. and L.M.C.; methodology, S.K.B., F.C. and L.M.C.; software, S.K.B. and F.C.; validation, S.K.B. and L.M.C.; formal analysis, S.K.B. and F.C.; investigation, S.K.B. and F.C.; resources, L.M.C.; data curation, S.K.B. and F.C.; writing—original draft preparation, S.K.B. and F.C.; writing—review and editing, L.M.C.; visualization, S.K.B. and F.C.; supervision, L.M.C.; project administration, L.M.C.; funding acquisition, L.M.C.

Funding: The authors gratefully acknowledge funding for this work from the G. Unger Vetlesen Foundation.

Acknowledgments: The authors thank Robert Davis and Jiahan Xie for equipment access and technical guidance. The authors also thank Drew Boland, Julia Davatzes, Curtis Davis, Cameron McCarty, Gabrielle Schleppenbach, and Ellie Wood for assistance with laboratory experiments.

Conflicts of Interest: The authors declare no conflict of interest. The funder had no role in the design of the study; in the collection, analyses, or interpretation of data; in the writing of the manuscript; or in the decision to publish the results.

References

1. Bhutto, A.W.; Qureshi, K.; Abro, R.; Harijan, K.; Zhao, Z.; Bazmi, A.A.; Abbas, T.; Yu, G. Progress in the production of biomass-to-liquid biofuels to decarbonize the transport sector-prospects and challenges. *R. Soc. Chem. Adv.* **2016**, *6*, 32140–32170. [CrossRef]
2. Elliott, D.C.; Biller, P.; Ross, A.B.; Schmidt, A.J.; Jones, S.B. Hydrothermal liquefaction of biomass: Developments from batch to continuous process. *Bioresour. Technol.* **2014**, *178*, 147–156. [CrossRef] [PubMed]
3. Toor, S.S.; Rosendahl, L.; Rudolf, A. Hydrothermal liquefaction of biomass: A review of subcritical water technologies. *Energy* **2011**, *36*, 2328–2342. [CrossRef]
4. Bauer, S.; Reynolds, C.; Peng, S.; Colosi, L. Evaluating the water quality impacts of hydrothermal liquefaction: Assessment of carbon, nitrogen, and energy recovery impacts. *Bioresour. Technol. Rep.* **2018**, *2*, 115–120. [CrossRef]
5. Jena, U.; Vaidyanathan, N.; Chinnasamy, S.; Das, K.C. Evaluation of microalgae cultivation using recovered aqueous co-product from thermochemical liquefaction of algal biomass. *Bioresour. Technol.* **2011**, *102*, 3380–3387. [CrossRef] [PubMed]
6. Dominguez-Faus, R.; Powers, S.E.; Burken, J.G.; Alvarez, P.J. The water footprint of biofuels: A drink or drive issue? *Environ. Sci. Technol.* **2009**, *43*, 3005–3010. [CrossRef] [PubMed]
7. Pham, M.; Schideman, L.; Scott, J.; Rajagopalan, N.; Plewa, M.J. Chemical and biological characterization of wastewater generated from hydrothermal liquefaction of Spirulina. *Environ. Sci. Technol.* **2013**, *47*, 2131–2138. [CrossRef]
8. Garcia Alba, L.; Torri, C.; Samorì, C.; Van der Spek, J.; Fabbri, D.; Kersten, S.R.A.; Brilman, D.W.F. Hydrothermal treatment (HTT) of microalgae: Evaluation of the process as conversion method in an algae biorefinery concept. *Energy Fuels* **2012**, *26*, 642–657. [CrossRef]
9. Valdez, P.J.; Nelson, M.C.; Wang, H.Y.; Lin, X.N.; Savage, P.E. Hydrothermal liquefaction of *Nannochloropsis* sp.: Systematic study of process variables and analysis of the product. *Biomass Bioenergy* **2012**, *46*, 317–331. [CrossRef]
10. Orfield, N.D.; Fang, A.J.; Valdez, P.J.; Nelson, M.C.; Savage, P.E.; Lin, X.N.; Keoleian, G.A. Life cycle design of an algal biorefinery featuring hydrothermal liquefaction: Effect of reaction conditions and an alternative pathway including microbial regrowth. *ACS Sustain. Chem. Eng.* **2014**, *2*, 867–874. [CrossRef]
11. Gai, C.; Zhang, Y.; Chen, W.T.; Zhou, Y.; Schideman, L.; Zhang, P.; Tommaso, G.; Kuo, C.T.; Dong, Y. Characterization of aqueous phase from the hydrothermal liquefaction of *Chlorella pyrenoidosa*. *Bioresour. Technol.* **2015**, *184*, 328–335. [CrossRef] [PubMed]
12. Tommaso, G.; Chen, W.; Li, P.; Schideman, L.; Zhang, Y. Chemical characterization and anaerobic biodegradability of hydrothermal liquefaction aqueous products from mixed-culture wastewater algae. *Bioresour. Technol.* **2015**, *178*, 139–146. [CrossRef] [PubMed]
13. Clarens, A.F.; Resurreccion, E.P.; White, M.A.; Colosi, L.M. Environmental life cycle comparison of algae to other bioenergy feedstocks. *Environ. Sci. Technol.* **2010**, *44*, 1813–1819. [CrossRef] [PubMed]
14. Kataki, S.; West, H.; Clarke, M.; Baruah, D.C. Phosphorus recovery as struvite from farm, municipal and industrial waste: Feedstock suitability, methods and pre-treatment. *Waste Manag.* **2016**, *49*, 437–454. [CrossRef] [PubMed]
15. Yetilmezsoy, K.; Fatih, I.; Emel, K.; Havva Melda, A. Feasibility of struvite recovery process for fertilizer industry: A study of financial and economic analysis. *J. Clean. Prod.* **2017**, *152*, 88–102. [CrossRef]
16. Ishii, S.K.L.; Boyer, T.H. Life cycle comparison of centralized wastewater treatment and urine source separation with struvite precipitation: Focus on urine nutrient management. *Water Res.* **2015**, *79*, 88–103. [CrossRef] [PubMed]
17. Capdevielle, A.; Sýkorová, E.; Biscans, B.; Béline, F.; Daumer, M. Optimization of struvite precipitation in synthetic biologically treated swine wastewater—Determination of the optimal process parameters. *J. Hazard. Mater.* **2013**, *244–245*, 357–369. [CrossRef]

18. Lahr, R.; Goetsch, H.; Haig, S.; Noe-Hays, A.; Love, N.; Aga, D.; Bott, C.; Foxman, B.; Jimenez, J.; Luo, T.; et al. Urine bacterial community convergence through fertilizer production: Storage, pasteurization, and struvite precipitation. *Environ. Sci. Technol.* **2016**, *50*, 11619–11626. [CrossRef]
19. Rahman, M.M.; Liu, Y.; Kwag, J.H.; Ra, C. Recovery of struvite from animal wastewater and its nutrient leaching loss in soil. *J. Hazard. Mater.* **2011**, *186*, 2026–2030. [CrossRef]
20. Yetilmezsoy, K.; Sapci-Zengin, Z. Recovery of ammonium nitrogen from the effluent of UASB treating poultry manure wastewater by MAP precipitation as a slow release fertilizer. *J. Hazard. Mater.* **2009**, *166*, 260–269. [CrossRef]
21. Çelen, I.; Buchanan, J.R.; Burns, R.T.; Robinson, R.B.; Raman, D.R. Using chemical equilibrium model to predict amendments required to precipitate phosphorus as struvite in liquid swine manure. *Water Res.* **2007**, *41*, 1689–1696. [CrossRef] [PubMed]
22. APHA. *Standard Methods for the Examination of Water and Waste Water*, 22nd ed.; American Public Health Association: Washington, DC, USA, 2013; Available online: http://www.standardmethods.org/PDF/22nd_Ed_Errata_12_16_13.pdf (accessed on 10 February 2016).
23. Xu, D.; Savage, P.E. Characterization of biocrudes recovered with and without solvent after hydrothermal liquefaction of algae. *Algal Res. Part A* **2014**, *6*, 1–7. [CrossRef]
24. Environmental Protection Agency (EPA). *Method 300.1: Determination of Inorganic Anions in Drinking Water by Ion Chromatography*; USEPA, Office of Water: Washington, DC, USA, 1997.
25. Gustafsson, J.P. Visual MINTEQ Ver. 3.1. KTH, Sweden. 2013. Available online: https://vminteq.lwr.kth.se/ (accessed on 20 March 2017).
26. Uysal, A.; Yilzmazel, Y.D.; Demirer, G.N. The determination of fertilizer quality of the formed struvite from effluent of a sewage sludge anaerobic digester. *J. Hazard. Mater.* **2012**, *181*, 248–254. [CrossRef] [PubMed]
27. Tchobanoglous, G.; Burton, F.; Stensel, H.D. *Wastewater Engineering: Treatment and Reuse*, 4th ed.; McGraw-Hill Higher Education: New York, NY, USA, 2003.
28. Connelly, E.B.; Colosi, L.M.; Clarens, A.F.; Lambert, J.H. Life cycle assessment of biofuels from algae hydrothermal liquefaction: The upstream and downstream factors affecting regulatory compliance. *Energy Fuels* **2015**, *29*, 1653–1661. [CrossRef]
29. Vardon, D.; Sharma, B.; Blazina, G.; Rajagopalan, K.; Strathmann, T. Thermochemical conversion of raw and defatted algal biomass via hydrothermal liquefaction and slow pyrolysis. *Bioresour. Technol.* **2012**, *109*, 178–187. [CrossRef]
30. Sawayama, S.; Minowa, T.; Yokoyama, S.-Y. Possibility of energy production and CO_2 mitigation by thermochemical liquefaction of microalgae. *Biomass Bioenergy* **1999**, *17*, 33–39. [CrossRef]
31. Jaffer, Y.; Clark, T.A.; Pearce, P.; Parsons, S.A. Potential phosphorus recovery by struvite formation. *Water Res.* **2002**, *36*, 1834–1842. [CrossRef]
32. Tao, W.; Fattah, K.P.; Huchzermeier, M.P. Struvite recovery from anaerobically digested dairy manure: A review of application potential and hindrances. *J. Environ. Manag.* **2016**, *169*, 46–57. [CrossRef]
33. Ali, M.I.; Schneider, P.A.; Hudson, N. Thermodynamics and solution chemistry of struvite. *J. Indian Inst. Sci.* **2005**, *85*, 141–149.
34. Jia, G.; Zhang, H.; Krampe, J.; Muster, T.; Gao, B.; Zhu, N.; Jin, B. Applying a chemical equilibrium model for optimizing struvite precipitation for ammonium recovery from anaerobic digester effluent. *J. Clean. Prod.* **2017**, *147*, 297–305. [CrossRef]
35. Molinos-Senante, M.; Hernández-Sancho, F.; Sala-Garrido, R.; Garrido-Baserba, M. Economic feasibility study for phosphorus recovery processes. *Ambio* **2010**, *40*, 408–416. [CrossRef]

Article

Extraction Behavior of Different Conditioned S. Rubescens

Michael Kröger [1,*], Marco Klemm [1] and Michael Nelles [1,2]

1 Deutsches Biomasseforschungszentrum gemeinnützige GmbH, Biorefineries Department, Torgauer Straße 116, 04347 Leipzig, Germany; marco.klemm@dbfz.de (M.K.); michael.nelles@uni-rostock.de (M.N.)

2 Faculty of Agricultural and Environmental Sciences, Chair of Waste Management, University of Rostock, Justus-von-Liebig-Weg 6, 18059 Rostock, Germany

* Correspondence: michael.kroeger@dbfz.de; Tel.: +49-341-2434-432

Received: 14 January 2019; Accepted: 3 April 2019; Published: 8 April 2019

Abstract: Microalgae utilized for experiments are often not produced by the researchers that are doing experiments with them. The microalgae are made storable through thermal or freeze-drying by the producer. In an industrial scaled process, because of efficiency reasons, microalgae would not be dried, but processed directly. With that, the question is, if drying already could change the composition or structure that much, that a process scaled up from laboratory to productive scale with fresh microalgae would be less efficient or even would not work at all. The effect of freeze drying on the extraction behavior for the species *Scenedesmus rubescens* was investigated. It was obtained in freeze-dried condition and again was delivered in fresh state. The utilized microalgae were extracted with n-hexane, without and with different pretreatments (acidic hydrolysis and hydrothermal carbonization) to examine the differences in the yields. In conclusion, it was demonstrated that freeze drying harms the cell wall and therefore this process already influences the quantity of extracted lipids. Depending on the harshness of the treatment process for cell wall disruption this might influence the extracted yield when the algae are not freeze-dried. The quality of the extracted lipids does not change when freeze-dried.

Keywords: microalgal oils; microalgae extraction; lipids; biofuel; cell wall disruption; hydrothermal disintegration; freeze-dried; fresh harvested; *Scenedesmus rubescens*

---

1. Introduction

Very often, the microalgae utilized for experiments are not produced by the researchers that are doing the experiments. The microalgae are made storable by the producer, which means they are harvested, centrifuged and heat or freeze-dried. This makes handling and timing for the consumer much easier. But these pretreatment steps are usually very energy intensive and especially the drying would not be conducted when working in an industrial process where microalgae would be produced as a bulk material and afterwards processed further, for example in an extraction process. For saving costs, in such an industrialized process the freshly harvested microalgae would be fed directly into the process. As it is likely, that beforehand in the R&D stage, freeze-dried microalgae were obtained and utilized, the question is, if the pretreatment (freeze drying) already did changes to the structure that makes the industrial process with fresh microalgae not that efficient or even does not work at all?

In the literature, several papers can be found that deal with the question of cell wall disruption because of an inevitable treatment like harvesting or some kind of dewatering or drying.

Falco et al. [1] analyzed spray dried *Spirulina platensis* with scanning electron microscopy (SEM) and showed that their morphology comprised of globular particles with an average size of several micrometers. The form of an intact *Spirulina (Arthrospira)* cell is helical/spiral. Therefore, the cell is already destroyed by the drying process. This is likely to happen with any process which applies mechanical pressure on a *Spirulina (Arthrospira)* cell.

Although a high disintegration grade for *Arthrospira* is already supposed to be reached by freeze drying and direct extraction, (i.e., a high fraction of lipids can be extracted without the need for a disintegration step), for other species this might be different [2,3]. In [4] *Chlorella* and *Scenedesmus* species needed a harsher method such as acidic hydrolysis and hydrothermal treatment to break up the cell wall structure and enhance the extraction yield.

Bohutskyi et al. [5] investigated the influence of harvesting techniques and storage conditions of several microalgae species and states that microalgae with a polysaccharide cell wall are less sensitive to rupture compared to those without it. In [6] a cyanobacterium *Aphanothece microscopica Nägeli* was dried at temperatures ranging from 40 °C to 60 °C at different thickness layers. The highest protein and lipid yields were achieved with 60 °C and a layer thickness of 7 mm (in comparison to 5 mm).

In [7] two diatoms (*Chaetoceros* sp. and *Phaeodactylum tricornutum*) were investigated towards their chemical composition after different preservation techniques. Direct freezing of the concentrated medium seemed the most reliable technique for the preservation of these microalgae. The proximate composition concerning proteins and lipids did only change slightly, even after a longer period (2 month). For air (30 °C) and freeze-dried algae these preservation techniques resulted in high percentage, around 70%, losses of lipids directly after treatment and also lost further organic components due to oxidation or bacterial activity after 2 months of storage. Protein content and profile did not change significantly. In contrast, Morist et al. [8] investigated treatment methods for *Spirulina platensis* pasteurization, spray-drying, and freeze-drying. Again, the protein content and profile did not change significantly, but the content of five different fatty acids was also measured. In this case the fatty acid content did not change to this high degree between the different treatment processes. The paper concluded that the freeze-drying method is the most recommended method because of almost no product degradation and higher biomass quality.

In this publication we investigate the effect of freeze drying for the species *Scenedesmus rubescens*. It was obtained in freeze-dried condition and, while doing experiments for another purpose, we had the chance to obtain algae from the same species from the same producer (IGV GmbH) in fresh state, still in the growth medium. The fresh microalgae were used as is (only centrifuged to lower the water content), but also for comparison, this charge of microalgae was freeze-dried and again processed like the original freeze-dried example. With this, it was possible to directly compare if it made a difference for the pretreatment and extraction if this species was freeze-dried or not. As already conducted in [4], the utilized microalgae were extracted with n-hexane, with and without different pretreatments (acidic hydrolysis and hydrothermal carbonization) to examine the differences in the yield and thereby give an answer on the consequences of making microalgae storable by freeze drying.

2. Materials and Methods

2.1. Material

The species used in this study; *Scenedesmus rubescens (S. rubescens)*, was provided kindly by the IGV GmbH, Germany. *S. rubescens* is a green microalga which grow in freshwater. About 2 kg of it was obtained. The microalgae was freeze-dried after harvesting. It was conducted at IGV GmbH. The freeze-dried algae had a residual water content of approximately 6 wt.%. The microalgae can be stored like this at room temperature for several months. The second (fresh) charge of *S. rubescens* was also obtained from IGV GmbH and was tested directly after centrifugation from the medium. Characteristics resulting from morphology should be the same, but because of seasonal differences and possibly different bioreactor set-ups between the freeze-dried and fresh obtained microalgae the absolute measures of the compounds may vary.

2.2. Freeze Drying

As the fresh obtained *S. rubescenes* from IGV GmbH should be compared to the ones obtained that were already freeze-dried by the producer, freeze drying also was conducted in our own lab. For this work, this was conducted with a special freeze dryer (Alpha 1-2 LDplus, Christ). For this, 100 g of the centrifugate of fresh *S. rubescens* was weighed in a crystallizing dish. The dish was covered by a watch glass and freeze-dried for 72 h until it was a constant mass. The residue was pulverized in a mortar prior to headspace-Karl-Fisher-Titration determining 8.5 wt.% residual water.

2.3. Properties

The raw material was dried in a Laboratory Drying Oven at 105 °C for 24 h to determine the water content [4]. The water content of the hydrothermal carbonization (HTC) product was determined by headspace-Karl-Fisher-Titration (Aqua 40.0, Elektrochemie Halle GmbH) based on DIN EN 14774-1 [4]. Carbon, nitrogen, hydrogen, and sulphur content were measured according to standard methods (DIN EN 15104) using an Elementar Vario Macro Cube (Elementar Analysensysteme GmbH, Hanau, Germany) [4]. C, H, N, and S were reported in weight percentage on a dry basis and higher heating value (HHV) in kJ/kg on dry basis [4]. Determination of the HHV was done by the CEN method (DIN EN 14918) using a Parr oxygen bomb calorimeter 6400 (Parr Instrument (Germany, Frankfurt)) [4].

2.4. Determination of Fatty Acid Profile by GC-MS

Analysis was performed on a 7890A gas chromatograph coupled with a mass-selective detector (5975C; Agilent Technologies, USA) [4]. The injection mode was split 1:15. A deactivated liner was inserted into the injection chamber, which was kept at 300 °C 1 μL of the sample was injected for separation by an ionic liquid phase (1,12-Di(tripropylphosphonium)dodecane bis (trifluoromethylsulfonyl) imide; SPB-IL 60 30 m × 0.25 mm x 0.20 μm, Sigma Aldrich) [4]. The gas chromatography (GC) system was operated in programmed-temperature mode: initial temperature 160 °C, first linear ramp 1 °C min^{-1} until 165 °C, 15 min hold, second linear ramp 5 °C min^{-1} until 250 °C final temperature, 1 min hold. Data acquisition was performed on the mass-selective detector in scan mode (40–450 amu) [4]. Due to changes in laboratory work flow a part of the measurements were done on a Polyethylenglycole phase (HP-INNOWax, 30 m × 0.25 mm × 0.25 μm; Agilent Technologies) The injection mode was split 1:15, the deactivated liner was kept at a temperature of 260 °C [4]. When using the HP-INNOWax-column, the GC system was operated in programmed-temperature mode as well: initial temperature 150 °C, 0.5 min hold, linear ramp 4 °C min^{-1} until 260 °C final temperature, 7 min hold. Data acquisition was performed on the mass-selective detector in scan mode (40–380 amu) [4]. An external calibration was done for both of the capillary columns separately [4].

Samples for GC-MS determination was prepared as follows: 100 mg of the extracted algae oil were dissolved in 5 ml tert-butyl methyl ether (MTBE) [4]. 100 μL of the MTBE solution was transferred to a GC-Vial equipped with a 0.2 ml micro-insert [4]. 50 μL trimethylsulfonium hydroxide (TMSH) were added to this solution. The mixture was shaken shortly and measured immediately [4]. The formed fatty acid methyl esters (FAME) were identified by mass spectral data (National Institute of Standards and Technology (NIST) 2008 Mass Spectral Library) and matching of retention times with FAME standard substances [4]. Quantitative determination was carried out by serial dilution of Grain Fatty Acids Methyl Ester Mix (Sigma Aldrich, Taufkirchen Germany) and applying an external linear calibration function [4].

2.5. Acidic Hydrolysis

For comparison and as an example for possible disintegration methods, microalgae were hydrolyzed by hydrochloric acid prior to the solvent extraction. In accordance to DIN 10342:1992-09 approximately 18 g of the homogenized algae sample, 135 mL water and 65 mL hydrochloric acid (37%) were stirred thoroughly and heated slowly until boiling [4]. Smooth boiling was maintained for 1 hour, followed by filtration and repeated rinsing of the residue and filter paper with distilled water [4]. The filter residue was dried at 105 °C for 24 h and weighed, yielding 61% hydrolyzed product.

2.6. Hydrothermal Treatment

The hydrothermal experiments were conducted in a 500 ml batch autoclave (Berghof, Highpreactor BR-500) with data logger, magnetic agitator and BTC 3000 temperature controller [4]. Temperature and pressure were recorded online. The system was stirred with 100 rpm to increase the heat transfer and avoid hot spots. Reaction temperature was typically 200 °C [4]. Heating time was 90 min (2 K/min rate) Holding time was in general 120 min [4]. Concentration of algae was 20 wt.%, as the algae had a water content of approximately 5.5 to 7 wt.% although being freeze-dried, the corrected dry weight was considered when calculating the yields of HTC char and lipids [4]. The cooling was carried out without external cooling over a period of several hours. After cooling, the resulting solid matter was separated by means of vacuum filtration and dried at 105 °C until a constant weight was attained [4]. The gaseous phase was expanded via a gas sampling valve. The volume of the gaseous phase was measured. As the produced gas mainly consisted of CO_2 and the volume produced is low in general at these process conditions [9] a further examination of the gaseous phase was neglected [4].

2.7. Extraction of Algae Oil

Extraction of the algae oil was done by Soxhlet extraction of the algae sample (freeze-dried or hydrolyzed) or the HTC product [4]. The solid was pulverized and homogenized in a porcelain mortar. The sample was transferred to a fat free extraction thimble and extracted continuously for 7 h [4]. As extraction agent n-hexane was used, as hexane extraction is presently seen as the most economical method for algae extraction [10].

All experiments were conducted only twice because of the limited amount of algae. It was not possible to obtain more microalgae of the same charge [4]. More repetitions of the experiments by using smaller amounts for each run would have led to less products for analyzing and less data. For all educts, intermediates and products, elemental composition, heating value, and, if possible, ash content were analyzed [4]. The extracted lipid yields were measured and put in relation to the input material [4].

3. Results and Discussion

The elemental composition of the microalgae without any treatment (Table 1) is very similar. It shows that there should only be minor differences in the composition of the different charges of *S. rubescens*.

Table 1. Elemental composition of the algae species and products.

Algae Species	C	H	S	N	HHV	H_2O
	wt.% (dw)	wt.% (dw)	wt.% (dw)	wt.% (dw)	kJ/kg (dw)	wt.% (dw)
S. rubescens						
Algae (freeze-dried)	48.40	8.61	0.37	9.30	21,670	5.62
HTC char	54.20	5.39	0.56	11.30	25,100	
Lipid from direct extraction	77.30	11.10	0.32	0.83	39,310	
Lipid HTC pretreated extraction	76.10	9.84	0.34	0.79	38,960	
Fresh *S. rubescens*						
Algae fresh (but freeze-dried)	50.80	8.98	0.44	9.05	22,290	8.49
HTC product from fresh S. rubescens	58.90	6.40	0.79	11.50	26,690	
Lipid from direct extraction of heat dried algae	74.50	10.40	0.37	0.91		
Lipid from direct extraction of freeze-dried (fresh) algae	74.70	10.50	0.33	0.31	37,780	
Lipid HTC pretreated extraction	72.00	10.60	0.34	0.91	38,880	

The HTC char yield of the fresh *S. rubescens* was between 45.5 and 55.9 wt.%. For freeze-dried *S. rubescens* it was between 53.6 and 54.9 wt.%. For the HTC char the carbon and hydrogen content was a bit higher for the fresh algae. For the directly extracted lipids, it was the original freeze-dried algae which had a higher carbon and hydrogen content. As this was only around 3 wt.% and 0.7 wt.% respectively, its questionable if this can be neglected.

The HTC pretreated extracts have differences in the carbon content, but the HHV was again already the same. The elemental composition of the freeze-dried *S. rubescens* and its products char and lipid was comparable to the values for the fresh *S. rubescens* algae and products. Therefore, in terms of the elemental composition the process of freeze drying in this work does not seem to have an influence.

The lipid yields achieved with the freeze-dried and fresh *S. rubescens* under the different pretreatments are given in Figure 1. Direct extraction of the freeze-dried *S. rubescens* gave the lowest yield with 55 g/kg. The acidic treatment gave an approximately 81% higher yield (100 g/kg). The yield after HTC treatment was 63 g/kg. For the fresh *S. rubescens*, again direct extraction gave the lowest yield. Freeze drying gave a yield of around 102 g/kg which was comparable to the yield after HTC treatment with 103 g/kg. The acidic treatment gave an approximately 80% higher yield (180 g/kg).

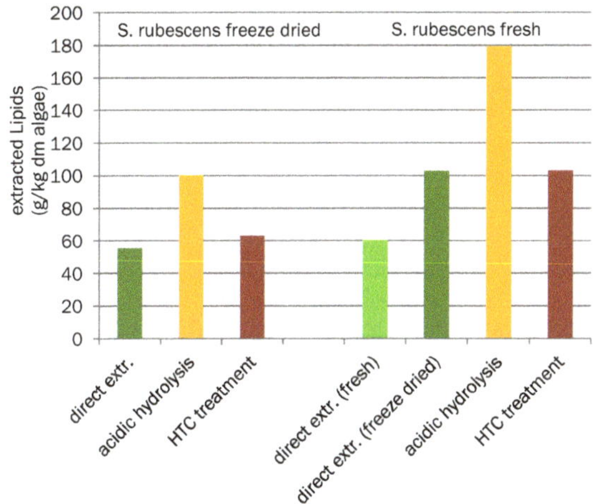

Figure 1. Extraction yields for fresh and freeze-dried *S. rubescens* at different pretreatments.

The lipid yields extracted from the fresh *S. rubescens* did vary in absolute measures from the yields of the freeze-dried *S. rubescens*. The reason for this can be found in seasonal differences and different bioreactor setups. Therefore, the absolute measures of the two charges can not be compared.

Nonetheless, as the species and therefore the morphology of the two charges is the same, the ratio of the different pretreatments is proportional for the two samples. The utilization of the fresh *S. rubescens* shows that mere freeze drying disintegrates the algae and elevates the extracted lipid fraction. For both samples (fresh and originally freeze-dried) the lipid yield was elevated by acidic hydrolysis at a similar ratio (74 wt.% and 81 wt.% respectively). The yield decreases for both fresh and freeze-dried *S. rubescens* when hydrothermally treated.

The GC-MS analysis of the Fatty Acids (FA) (Figure 2) generally showed a good correlation between originally freeze-dried and fresh *S. rubescens*. The sum of the direct extracted lipids was about 10 wt.% lower for the freeze-dried *S. rubescens*, but this is not due to one specific FA.

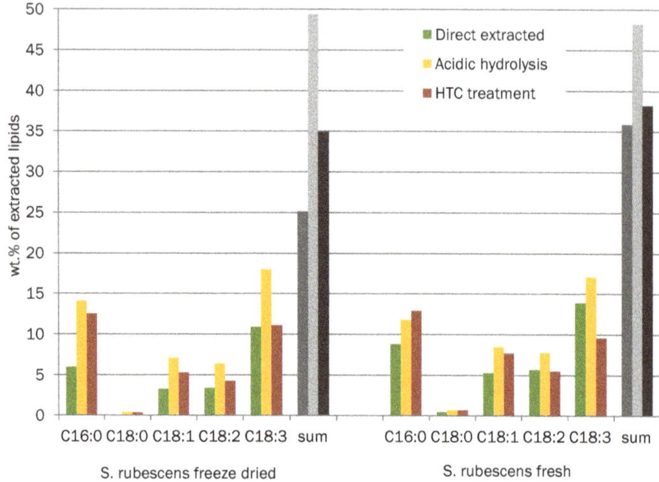

Figure 2. wt.% of main fatty acids and percentage of recognized fraction of *S. rubescens*.

In Figure 3 the percental fraction of recognized FAs was displayed for the fresh and freeze-dried *S. rubescens*. C16:0 increases from direct to acidic hydrolysis to HTC extraction for the freeze-dried example. For the fresh *S. rubescens* only the HTC extracted yield was higher. For C18:0 and C18:1 the increase from direct to acidic hydrolysis to HTC extraction was similar. For C18:2 the differences between the yields of the extraction methods are low, but the decrease from direct to HTC extraction can be seen for both examples. For the C18:3 FA the percental yield decreases for both algal examples. The decrease of the C18:3 FA might be because of a concentration change when breaking the cell wall and diluting other kinds of lipids. On the other hand, one or more of the double bonds could be opened by a hydrolysis or oxygenation reaction. A decrease of C18:3 FA with increase in temperature was also determined in [11]. The increase of the C16:0 FA might be due to a higher content of C16:0 FA in the cell wall.

Thereby, the proportional changes in the FA composition for the different methods show a good correlation between the freeze-dried and fresh *S. rubescens*. The increasing and decreasing of the yields from direct to acidic hydrolysis to HTC extraction was generally quite similar for all fatty acids displayed.

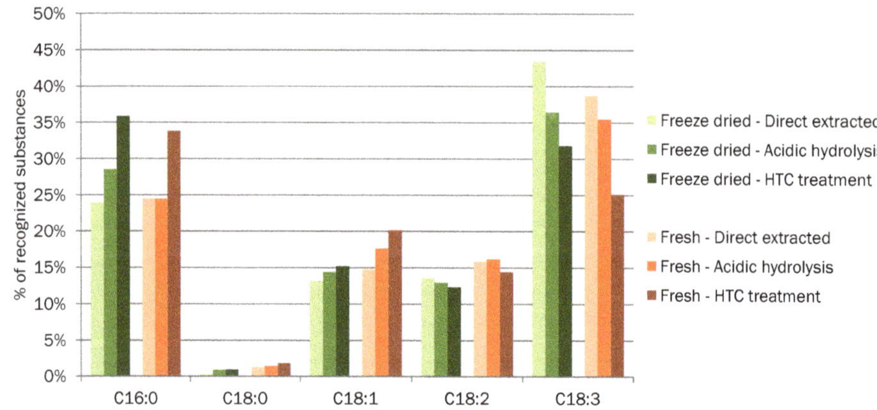

Figure 3. Main fatty acid percentage of the sum of recognized fraction for *S. rubescens*.

4. Conclusions

Concerning the question if using fresh or freeze-dried material for laboratory experiments matters, it has to be stated, that there is a difference in the results found in the conducted experiments for this work. The freeze drying of the fresh *S. rubescens* gave a higher yield of lipids then the direct extraction of the fresh ones. Therefore, there must have been some kind of disruption of the cell wall, which has to be taken into account when scaling up the considered process. For the lipid extracts it can be stated that freeze drying does not thoroughly alter the fatty acid composition and thereby should not lead to problems in process scale up.

Author Contributions: All authors contributed substantially to all aspects of this article.

Funding: This research received no external funding.

Conflicts of Interest: The authors declare no conflict of interest.

References

1. Falco, C.; Sevilla, M.; White, R.J.; Rothe, R.; Titirici, M.-M. Renewable Nitrogen-Doped Hydrothermal Carbons Derived from Microalgae. *Chem. Sus. Chem.* **2012**, *5*, 1834–1840. [CrossRef] [PubMed]
2. Northcote, D.; Goulding, K.; Horne, R. The chemical composition and structure of the cell wall of Chlorella pyrenoidosa. *Biochem. J.* **1958**, *70*, 391. [CrossRef] [PubMed]
3. Venkataraman, L.; Shivashankar, S. Studies on the extractability of proteins from the alga Scenedesmus acutus. *Algol. Stud. /Arch. Für Hydrobiol. Suppl.* **1979**, *22*, 114–126.
4. Kröger, M.; Klemm, M.; Nelles, M. Hydrothermal Disintegration and Extraction of Different Microalgae Species. *Energies* **2018**, *11*, 450. [CrossRef]
5. Bohutskyi, P.; Betenbaugh, M.J.; Bouwer, E.J. The effects of alternative pretreatment strategies on anaerobic digestion and methane production from different algal strains. *Bioresour. Technol.* **2014**, *155*, 366–372. [CrossRef] [PubMed]
6. Zepka, L.Q.; Jacob-Lopes, E.; Goldbeck, R.; Queiroz, M.I. Production and biochemical profile of the microalgae Aphanothece microscopica Nägeli submitted to different drying conditions. *Chem. Eng. Process. Process Intensif.* **2008**, *47*, 1305–1310. [CrossRef]
7. Esquivel, B.C.; Lobina, D.V.; Sandoval, F.C. The biochemical composition of two diatoms after different preservation techniques. *Comp. Biochem. Physiol. Part B Comp. Biochem.* **1993**, *105*, 369–373. [CrossRef]
8. Morist, A.; Montesinos, J.L.; Cusidó, J.A.; Gòdia, F. Recovery and treatment of Spirulina platensis cells cultured in a continuous photobioreactor to be used as food. *Process Biochem.* **2001**, *37*, 535–547. [CrossRef]
9. Hoekman, S.K.; Broch, A.; Robbins, C. Hydrothermal Carbonization (HTC) of Lignocellulosic Biomass. *Energy Fuels* **2011**, *25*, 1802–1810. [CrossRef]
10. Kannan, D.C.; Pattarkine, V.M. Recovery of Lipids from Algae. In *Algal Biorefineries*; Bajpai, R., Prokop, A., Zappi, M., Eds.; Springer: Cham, The Netherlands, 2014; pp. 297–310. ISBN 978-94-007-7493-3.
11. Biller, P.; Friedman, C.; Ross, A.B. Hydrothermal microwave processing of microalgae as a pre-treatment and extraction technique for bio-fuels and bio-products. *Bioresour. Technol.* **2013**, *136*, 188–195. [CrossRef] [PubMed]

Article

Hydrothermal Carbonization of Various Paper Mill Sludges: An Observation of Solid Fuel Properties

Nepu Saha [1,2], Akbar Saba [1,3], Pretom Saha [1,3], Kyle McGaughy [1,2], Diana Franqui-Villanueva [4], William J. Orts [4], William M. Hart-Cooper [4] and M. Toufiq Reza [1,2,*]

1 Institute for Sustainable Energy and the Environment, Ohio University, Athens, OH 45701, USA; ns671517@ohio.edu (N.S.); as030616@ohio.edu (A.S.); ps332316@ohio.edu (P.S.); km067812@ohio.edu (K.M.)
2 Department of Chemical and Biomolecular Engineering, Ohio University, Athens, OH 45701, USA
3 Department of Mechanical Engineering, Ohio University, Athens, OH 45701, USA
4 USDA-PWA, ARS, WRRC, BCE, 800 Buchanan Street, Albany, CA 94710, USA; diana.franqui@ars.usda.gov (D.F.-V.); bill.orts@ars.usda.gov (W.J.O.); william.hart-cooper@ars.usda.gov (W.M.H.-C.)
* Correspondence: reza@ohio.edu; Tel.: +1-740-593-1506

Received: 31 January 2019; Accepted: 27 February 2019; Published: 5 March 2019

Abstract: Each year the pulp and paper industries generate enormous amounts of effluent treatment sludge. The sludge is made up of various fractions including primary, secondary, deinked, fiber rejects sludge, etc. The goal of this study was to evaluate the fuel properties of the hydrochars produced from various types of paper mill sludges (PMS) at 180 °C, 220 °C, and 260 °C. The hydrochars, as well as the raw feedstocks, were characterized by means of ultimate analysis, proximate analysis, moisture, ash, lignin, sugar, and higher heating value (HHV_{daf}) measurements. Finally, combustion indices of selected hydrochars were evaluated and compared with bituminous coal. The results showed that HHV_{daf} of hydrochar produced at 260 °C varied between 11.4 MJ/kg and 31.5 MJ/kg depending on the feedstock. This implies that the fuel application of hydrochar produced from PMS depends on the quality of feedstocks rather than the hydrothermal carbonization (HTC) temperature. The combustion indices also showed that when hydrochars are co-combusted with coal, they have similar combustion indices to that of coal alone. However, based on the energy and ash contents in the produced hydrochars, Primary and Secondary Sludge (PPS_2) could be a viable option for co-combustion with coal in an existing coal-fired power plant.

Keywords: primary sludge; secondary sludge; deinked sludge; fiber rejects; hydrochar; combustion indices; fuel properties

1. Introduction

The world energy demand continues to grow with the increase of populations. The energy demand is projected to reach up to 42 quadrillion British Thermal Unit (BTU) in 2040 to fulfill the needs of 10 billion people [1,2]. In order to ensure this energy supply is provided in a sustainable manner, fossil fuel industries need to begin integrating with renewable energy sources. Coal power plants, one of the major electricity producers in the U.S., are still expected to provide 17.6% more electricity in 2040 compared to today [2]. This sustainability problem provides extraordinary potential for coal-like fuels generated from renewable resources. Wet waste biomass, such as paper mill sludges (PMS), is one of the largest and most abundant sources of renewable energy [3]. In fact, around 16.5 million dry tons of PMS have been produced annually in the U.S. These wastes have been landfilled, land applied, incinerated, or used in other ways (e.g., filler, fibers, composite etc.) [4]. Due to the strict landfill policies and shortage of lands, incineration is becoming the preferred treatment method despite having incineration costs of around $13–15 per dry ton [5]. Co-firing PMS with coal in existing coal-fired power plants may reduce the carbon footprint for the power plant and it may reduce the cost of waste disposal for the paper mill by generating a sellable product rather than waste [6,7]. However, viable co-firing requires a feedstock of consistent quality with higher energy density and combustion characteristics compatible with coal.

A wide variety of PMS are generated in the pulp and paper mills [8–10]. PMS can be categorized into four broad sections: (1) primary sludge (PS), which is produced in the primary clarifier and contains mainly fines and fillers [11]; (2) de-inking paper sludge (DPS), which is produced during the floatation process and is very common for recycled papers [12]; (3) secondary sludge (SS) or biological sludge, which is the effluent of the microbial wastewater treatment system and is made of microbial masses [10]; and (4) combined primary and secondary sludge (PSS), which is a mixture of primary and secondary sludges from the same plant [13]. Besides these basic four categories, there could be preliminary fiber rejects (PFR) containing preliminary impurities such as metals, inorganics, plastics, tree branches, etc. and pre-thickened sludge (PTS), which is primary sludge prior to the dewatering step [14]. The chemical composition of the PMS varies significantly among these categories [8]. For instance, PS is more fibrous and easier to dewater when compared to SS and PSS [11]. DPS contains pigments, fillers, additives, and coatings in addition to the fiber particles [12]. Meanwhile, PTS contains high moisture (>98%) and fiber rejects (FR) contains very low moisture but also high ash content [10]. Regardless, all the PMS are wet, which prevent them from being applied directly for energy generation.

Hydrothermal carbonization (HTC) is an emerging technology for producing biofuels and upgrading solid fuels [15–18]. It conveniently uses residual moisture of the feedstocks as the reaction medium due to its solvent properties at high temperatures and pressures [19]. Therefore, wastes do not need to be dried prior to HTC, which is typically an energy intensive process. Subcritical water at around 200–260 °C has high ionic products and low dielectric constants, which allow water to be more reactive and cause it to behave similarly to non-polar solvents [20]. Therefore, HTC conditions allow for carbon based organic compounds (namely, hemicellulose and cellulose) to be hydrolyzed into monomers. These monomer fragments then undergo a series of reactions including dehydration, decarboxylation, condensation, and polymerization, which can result in products such as hydrochars and organic acids [15,16,21]. These reactions follow separate kinetics and can be catalyzed by acids, bases, and inorganics [22–24]. As a result of these complex reaction rates, a wide range of residence times ranging from 5 min to 6 h have been reported for HTC [18,25,26]. HTC process results in a solid hydrochar that is quite hydrophobic, friable, and energy dense [27–29]. Although temperature is considered as the most important processing parameter for an HTC process, the types of feedstock (e.g., herbaceous, woody, municipal sludge, food waste, sewage sludge, PMS, etc.) and their chemical compositions also determine the chemical characteristics of hydrochar [30–32]. As a result, different optimum HTC conditions are reported in the literature for different feedstocks [33–36]. Although large-scale implementation of HTC is yet to established, recent techno-economic analyses have shown that HTC of biomass can be economically viable and environmentally sustainable [37–39].

HTC of PMS has been extensively studied in the last few years. For instance, Areeprasert et al. performed HTC of PSS in a batch reactor at 180–240 °C for 30 min to determine optimal HTC conditions [40–42]. Afterwards, HTC was performed at the optimized condition (197 °C and 30 min) in a small pilot scale [43]. Their results indicate that HTC of PSS is energetically favorable in the context of Japan. Furthermore, the same group have determined combustion characteristics and combustion kinetics of PSS hydrochar and reported that the major decomposition was attributed for cellulose [44]. Meanwhile, the HTC process liquid resulting from PSS was analyzed and recycled in order to reduce the emission footprint of the paper mill [45]. In order to understand the HTC reaction parameters, Makela et al. performed HTC of PSS from 180–260 °C for 30 min–5 h, both with and without homogeneous catalysts [26,46]. Their results indicate that dehydration, depolymerization, and decarboxylation are more prominent than polymerization and aromatization. Similar to the other biomass feedstocks, hydrochloric acids catalyzed HTC of PSS resulted in an energy yield increase of the hydrochar [46].

Literature has repeatedly used PSS or mixed sludge for HTC experiments, whereas the HTC of other PMS are rare and the differences in chemical compositions among the various PMS are likely to result in variations in hydrochars. Therefore, the goal of this study was to examine the fuel quality of solid hydrochars produced by various PMS at various HTC temperatures. The first objective of this study was to reveal whether the type of PMS feedstocks is more influential than the HTC temperature in terms of energy densification (ED). As the ultimate goal is to co-fire PMS hydrochar with coal in an existing coal-fired power plant, the second objective was to evaluate the combustion indices of the PMS hydrochars compared to the combustion indices of bituminous coal.

2. Materials and Methods

2.1. Materials

Six different PMS were obtained from an industrial partner's different pulp and paper mill plants found throughout the USA specifically for this study. Table 1 displays the samples names and how they are referenced throughout the manuscript. Samples PS, DPS, PSS_1, PSS_2, and PFR are all from different plants; PTS was obtained from the same plant as PFR samples. Meanwhile, Clarion #5a (bituminous coal) was used for co-combustion studies. All samples were dried in an oven at 105 °C for 24 h and sealed in Ziploc bags until further use. The nomenclature of the samples and their moisture contents are shown below in Table 1.

Table 1. Sample nomenclature used throughout manuscript and their respective moisture contents.

Sludge Sample	ID	MC (%)
Primary Sludge	PS	60.1 ± 1.5
De-inked Paper Sludge	DPS	63.5 ± 0.5
Primary and Secondary Sludge 1	PSS_1	64.1 ± 1.0
Primary and Secondary Sludge 2	PSS_2	76.5 ± 1.0
Primary Sludge and Fiber Rejects	PFR	57.1 ± 1.2
Pre-thickened Sludge	PTS	98.7 ± 0.1

2.2. Hydrothermal Carbonization (HTC)

All HTC experiments were performed in a 600 mL Parr 452HC3 Hastelloy reactor (Parr Instrument Company, Moline, IL, USA). Reactor temperature was controlled with a Parr 4843 proportional–integral–derivative controller (PID) controller using a J-type thermocouple with an accuracy of ±3 °C. Reaction pressure was monitored as the reactor was a closed system, thus observed pressures were autogenous pressures in addition to the pressures exerted by produced gases. The reactor was loaded with a 1:9 dry sample to de-ionized (DI) water ratio. Once the reactor was sealed, the reactor stirrer was set to an RPM rate of 180 ± 2 rpm. The reactor mixtures were stirred from then until the end of the HTC process. The hydrochars were produced at reaction temperatures of 180 °C, 220 °C, and 260 °C. The reactor heating procedure was as follows: (1) the contents of the reactor were heated at approximately 8 °C min^{-1} until the set reaction temperature, (2) once at the target reaction temperature, the reactor temperature was held constant for a residence time of 30 min, (3) after the residence time, the reactor was submerged in an ice-water bath until it reached to 30 °C, which typically took around 10 min, and (4) reactor stirring was stopped once 30 °C was reached. Produced gases were vented in a fume hood and the process liquid was separated from the solid hydrochar via vacuum filtration and a Whatman 41 filter (20 µm). Hydrochar was washed with DI water until the pH of washed water reached the pH of DI water. The moist hydrochars were dried in an oven at 105 °C for 24 h, then stored in a Ziploc bag until further analysis.

2.3. Characterization of Hydrochars

Mass yields were calculated using Equations (1)–(3) and they show how much of the initial feedstock was converted into solid (hydrochar), liquid and gas.

$$\text{Solid Mass Yield } (\%) = \frac{\text{Mass of dry hydrochar}}{\text{Mass of untreated dry feedstock}} \times 100\% \quad (1)$$

$$\text{Liquid Mass Yield } (\%) = \frac{\text{Mass of liquid produced}}{\text{Mass of untreated dry feedstock}} \times 100\% \quad (2)$$

$$\text{Gas Mass Yield } (\%) = \frac{\text{Mass of gas produced}}{\text{Mass of untreated dry feedstock}} \times 100\% \quad (3)$$

Carbon, hydrogen, nitrogen, and sulfur content were determined using a Flash 2000 Organic Elemental Analyzer (Thermo Scientific, Grand Island, NY, USA). The oxygen content was calculated by subtracting carbon, hydrogen, nitrogen, sulfur, and ash percentages from 100%. Electrolytic copper and copper oxide were used for the analyzer's oxidation column and 2,5-Bis (5-tert-butyl-benzoxazol-2-yl) thiophene (BBOT) was used for calibration. Higher heating valued (HHV) were also determined for hydrochars using a Parr 6200 adiabatic oxygen-bomb calorimeter (Parr Instrument Company, Moline, IL, USA) calibrated with benzoic acid. Dry ash free HHV (HHV_{daf}) was calculated via Equation (4). Energy densification (ED) was calculated via Equation (5) to evaluate how HHV changes from the untreated feedstock to hydrochar.

$$HHV_{daf} = \frac{\text{HHV of the sample}}{1 - \text{Fraction of dry ash into the sample}} \quad (4)$$

$$ED = \frac{\text{HHV of dry hydrochar}}{\text{HHV of untreated dry feedstock}} \quad (5)$$

Lignin analysis was performed by the standard Technical Association of the Pulp and Paper Industry (TAPPI) method [47]. A modified method of Sequeira and Law [48] was used to quantify the sugar content. In this method, DI-water at 50 °C was used for making and blending a 5% solid slurry for one minute; three extractions were performed. Solid was filtered in a pressurized air filter (OFI Testing Equipment, model 140-31) by using Whatman no. 4 filter paper. The filtration process was operated just above atmospheric pressure to ensure liquor was extracted from the sludges or hydrochars. Final extraction was conducted in an autoclave (L30 cycle) to make sure no free sugar remained in the solid. HPLC (Agilent 1200 series) was used to analyze the sugar and sugar alcohols. HPLC was equipped with a binary pump, an HPX-87P (Bio-Rad) column, and a refractive index (RI) detector. Water was used as a mobile phase with a flowrate of 0.5 mL min^{-1}. The column was heated to 85 °C to enhance the separation and the RI detector was used at 55 °C. An external calibration was used to quantify the contents (lactose, glucose, xylose, galactose, arabinose, mannose, and fructose).

Volatile matter, fixed carbon, and ash content of samples were determined by thermogravimetric analyses (TGA) using a TGA Q500 (TA instruments, New Castle, DE, USA). Sample heat rate was set at 20 °C min^{-1} and all gas flowrates were set at 50 mL min^{-1}. A nitrogen atmosphere was used for proximate analysis. Samples were heated from 25 °C to 105 °C and held at 105 °C for 5 min, the temperature was then increased to 900 °C and then held for 5 min. Air was then introduced for 10 min to combust the remaining sample left at 900 °C. The mass loss under the nitrogen atmosphere at 105 °C was considered moisture. Mass loss between 105 °C and 900 °C were considered volatile matters. Mass remaining at the end of the combustion process was considered ash. Fixed carbon was determined by subtracting moisture, volatile and ash percentages from 100%.

Oxidative reactivity was performed to determine fuel properties and was performed similarly to the proximate analysis method described earlier using the same equipment. Similar heating rates, gas flows, and temperature increments were used, but air was the only gas used during the entire run. Combustion thermograms and derivative of the mass relative to temperature (DTG) curves were plotted as shown in the literature [49]; calculation of ignition (Equation (6)) and burnout (Equation (7)) indices are presented below [50]:

$$D_i = \frac{|DTG_{max}|}{t_p t_i} \tag{6}$$

$$D_B = \frac{|DTG_{max}|}{\Delta t_{1/2} t_p t_B} \tag{7}$$

where, DTG_{max} is the maximum rate of combustion, in magnitude, found in the combustion DTG curves; t_p is the time where DTG_{max} occurs; t_i and t_B are the time at which the ignition temperatures and burnout temperatures are obtained, respectively; and $\Delta t_{1/2}$ is the time when the combustion DTG value is one-half of the DTG_{max}. Lower values of D_i and D_B indicate an ideal ignition and burnout (i.e. overall combustion) behavior.

3. Results and Discussion

3.1. Product Distribution and Energy Densification

Table 2 shows the solid, liquid, and gas mass yields of with respect to different HTC temperatures. The initial solid to liquid ratio for each experiment was maintained at 1:9. With the increase of HTC temperature, liquid percentage increases and at the same time solid percentage decreases in the output. The increase of liquids resulted from the decomposition reactions that occur during HTC; mainly in the form of dehydration, polymerization, and aromatization reactions [21,51]. Although there was no gas added initially in the reactor, a small amount of gas (less than 1 g total) had been generated during HTC, especially at higher HTC temperatures. Decarboxylation of PMS polymers may resulted in the gaseous product [21,52]. Since the reactivity of biomass increases with the HTC temperature, lower solid mass yields were expected with the increase of HTC temperatures.

Table 2. Mass yields and energy densification of paper mill sludges (PMS) and their hydrochars ([a] Not applicable).

Sample	HTC Temperature (°C)	Solid Mass Yield (%)	Liquid Mass Yield (%)	Gas Mass Yield (%)	HHV_{daf} (MJ kg^{-1})	ED
PS	Raw	NA [a]	NA [a]	NA [a]	15.5 ± 0.0	NA [a]
	180	96.2 ± 2.0	3.8 ± 2.0	0.0 ± 0.0	15.1 ± 0.4	1.0 ± 0.0
	220	85.4 ± 2.0	11.8 ± 2.2	2.8 ± 0.2	15.3 ± 0.4	1.0 ± 0.0
	260	30.3 ± 1.1	58.1 ± 2.4	11.6 ± 1.4	22.8 ± 0.1	1.5 ± 0.0
DPS	Raw	NA [a]	NA [a]	NA [a]	13.0 ± 0.0	NA [a]
	180	91.7 ± 0.9	8.3 ± 0.9	0.0 ± 0.0	13.0 ± 0.2	1.0 ± 0.0
	220	84.4 ± 0.9	13.3 ± 1.4	2.3 ± 0.5	12.9 ± 0.1	1.0 ± 0.0
	260	58.4 ± 2.0	33.9 ± 1.3	7.2 ± 1.3	11.4 ± 0.7	0.9 ± 0.1
PSS_1	Raw	NA [a]	NA [a]	NA [a]	21.4 ± 0.1	NA [a]
	180	81.2 ± 0.3	18.3 ± 1.0	0.0 ± 0.0	21.7 ± 0.1	1.0 ± 0.0
	220	74.1 ± 0.0	22.2 ± 0.4	3.6 ± 0.5	21.7 ± 0.1	1.0 ± 0.0
	260	54.1 ± 2.0	38.8 ± 3.6	7.1 ± 1.6	27.4 ± 0.1	1.3 ± 0.0
PSS_2	Raw	NA [a]	NA [a]	NA [a]	19.6 ± 0.2	NA [a]
	180	90.9 ± 0.5	9.1 ± 0.5	0.0 ± 0.0	19.6 ± 0.1	1.0 ± 0.0
	220	78.9 ± 1.2	18.4 ± 1.2	2.7 ± 0.0	20.2 ± 0.1	1.0 ± 0.0
	260	41.1 ± 1.7	49.9 ± 2.1	9.0 ± 0.4	28.9 ± 0.1	1.5 ± 0.0
PFR	Raw	NA [a]	NA [a]	NA [a]	19.7 ± 0.2	NA [a]
	180	93.2 ± 2.1	6.8 ± 2.1	0.0 ± 0.0	18.5 ± 0.0	0.9 ± 0.0
	220	81.9 ± 1.2	15.6 ± 0.5	2.5 ± 0.7	19.3 ± 0.2	1.0 ± 0.0
	260	45.4 ± 1.2	42.9 ± 0.8	10.7 ± 0.4	25.2 ± 0.3	1.3 ± 0.0
PTS	Raw	NA [a]	NA [a]	NA [a]	17.1 ± 0.7	NA [a]
	180	87.5 ± 2.9	8.0 ± 0.8	0.0 ± 0.0	18.6 ± 0.5	1.1 ± 0.0
	220	69.8 ± 2.8	25.2 ± 1.4	5.0 ± 1.4	21.6 ± 0.1	1.3 ± 0.0
	260	43.8 ± 1.1	52.0 ± 4.6	4.1 ± 3.5	31.5 ± 3.7	1.8 ± 0.3
Clarion # 5a coal	Raw	NA [a]	NA [a]	NA [a]	30.5 ± 0.3	NA [a]

The variation of HTC product distribution resulting from different PMS feedstocks can be found in Table 2. Solid mass yields of HTC treated PSS_1 and PSS_2 considerably decreased as reaction temperature increased, reaching yields as low as 54.1% and 41.1%, respectively at 260 °C. Even though both are mixtures of primary and secondary sludges, their solid mass yields were significantly different. The reason could simply be the variation of ratios of primary to secondary sludges in the samples. This result shows that sludge feedstocks can vary in a manner that impacts HTC performance, even amongst the same category of sludges. The yields at 180 °C and 220 °C were quite high for both feedstocks, which may indicate that both sludges were rich in cellulose and lignin. As the cellulose and lignin react with water during HTC at approximately 220–230 °C and 255–265 °C, respectively, PMS remain mostly unreacted at 180 °C and 220 °C [52–55]. So far, researchers have mainly conducted their research on the mixed (primary and secondary) PMS and have found the solid mass yield in a range between 29 to 65% depending on the mixing ratio of primary to secondary sludges when the HTC was conducted at 260 °C [26,56].

Solid mass yield of HTC treated PS significantly decrease at the high reaction temperature, reaching a yield as low as 30.3% at 260 °C. However, minimal changes were observed at 180 and 220 °C, which indicate that the PS is rich in cellulosic material that did not completely react until 260 °C. Kim et al. found that the PS contains 58% cellulose [57]. The solid mass yield of PFR was similar to PS at low temperatures (180 and 220 °C), while solid mass yield of PFR was higher than the solid mass yield of PS at 260 °C. The reason for this higher solid mass yield for PFR could be the presence of higher lignocellulosic material in PFR compared to PS, as PFR contains tree branches and barks. These lignin-rich compounds breakdown slowly, and likely remain unreacted at these treatment conditions.

Since PTS is basically the PS prior to the dewatering step [14], the solid mass yield for PTS are expected to be similar to PS. However, a significant difference was observed between their solid mass yields, especially at 260 °C. The solid mass yield was 43.8% for PTS while it was 30.3% for PS. Although the yields were different, they fall within the literature values of PS (29 to 65%). This difference indicates that the PTS might have an even distribution of lignin and cellulosic compounds, as it seems similar to degradation from 180 °C to 220 °C to 260 °C. With PS, more significant difference occurs between 220 °C and 260 °C than between 180 °C and 220 °C. This indicates that the PS is composed of more hardy lignin compounds that break down at higher temperatures. These differences in solid mass yield with respect to temperature are consistent with previous literature, which has traditionally explained this observation as being the result of difference in feedstocks. A mild change in solid mass yield at 260 °C was observed for DPS, which was 58.4%. This higher solid mass yield, compared to the other sludges, may be due to either the sludge containing more inorganic components which remain unreacted during HTC, or the sludge containing more lignocellulosic fiber wastes which do not degrade as readily under these conditions.

It was observed from the mass yields that minimal carbonization happened to the PMS at 180 °C and 220 °C, even though cellulose degradation is expected to occur at 220 °C. Meanwhile, significant changes only happened to all PMS at 260 °C. This indicates that the cellulosic mass may be stabilized by the other lignin components in the PMS, requiring a higher temperature to initiate the HTC process. Mass yields alone, however, do not give a full account of the benefits of the HTC process. As the main focus of this study is the fuel properties of solid hydrochars, it is necessary to check additional fuel characteristics (i.e., HHV or ED) of all hydrochars to evaluate the more impactful hydrochars for further analysis.

How the energy is concentrated into the hydrochar with respect to its raw sample is known as energy density (ED). The dry ash-free HHV (HHV_{daf}) and ED, relative to the raw samples, are shown in Table 2. The HHV_{daf} of raw sludges varies from 13.0 to 21.4 MJ kg^{-1}, where DPS has the minimum, PSS_1 has the maximum, and the others fall in between. The HHV_{daf} of all sludges are lower than the HHV of bituminous coal (25.4–27.4 MJ kg^{-1}) [58,59]. The HHV_{daf} of hydrochars produced at 180 and 220 °C were similar to their individual raw samples. However, the HHV_{daf} of hydrochars produced at 260 °C were significantly higher than the corresponding raw samples, with the exception of DPS. Some of them are even higher than the HHV_{daf} of coal (e.g., hydrochar of PSS_1, PSS_2, and PTS produced at 260 °C). The HHVs of hemicellulose and cellulose have a range between 16.8–18.6 MJ kg^{-1} and break down around 180 °C and 220–230 °C [53,54,60]. On the other hand, lignin not only has a higher HHV (23.3–25.6 MJ/kg) but also has HTC initiation around 255–265 °C [52,55]. As a result, the component that has higher HHV stays in the hydrochar product at a higher HTC temperature. The only exception was observed in DPS, where the HHV decreases with the increase of HTC temperature. This decrease in HHV, which is unique to the DPS char, is due to the loss of carbon and relative increase in oxygen (both must be compared to the total mass loss of the sludge). The lack of carbonization could be due to the higher amount of inorganics present in the DPS, which is shown in its high ash content.

Table 2 shows that ED remains 1.0 for all hydrochars produced at 180 and 220 °C, which indicates that almost nothing favorable happened for PMS in terms of energy content at those HTC conditions. However, the energy contents by the hydrochars produced at 260 °C were 1.3 to 1.8 times higher than their raw samples except for DPS, which had ED lower than 1.0 at 260 °C. As such, the types of feedstocks are more significant than HTC reaction temperature for densifying the energy in PMS. As solid fuel properties of PMS are the main focus of this study and significant changes of ED happened only at HTC 260 °C, from now on, the characteristics of hydrochars produced at 260 °C will be discussed in the following sections.

3.2. Chemical Characteristics of PMS Hydrochars Produced at 260 °C

From the previous sections, it was observed that the solid mass yield decreases and ED generally increases with the HTC temperature. Therefore, it is important to evaluate how the organics and lignin in the hydrochars produced at 260 °C vary from their raw sludges, including their significant impact on ED. Table 3 presents the ultimate analysis including ash, lignin, and sugar (C_5 and C_6) content of all PMS and their corresponding hydrochars produced at 260 °C. Hydrochar of each sample has higher carbon content compared to its raw sample, except DPS. The carbon content increases from 34.6% to 44.3%, 38.4% to 40.5%, 42.8% to 62.2%, 39.8% to 57.6%, and 23.9% to 40.2% for PS, PSS_1, PSS_2, PFR, and PTS, respectively. However, this content decreases for DPS from 27.1% to 22.2%. The oxygen content, on the other hand, has a decreasing trend for all sludges from raw to its hydrochar. The oxygen content decreases from 44.4% to 18.5%, 37.0% to 29.6%, 30.3% to 16.0%, 45.8% to 25.0%, 34.6% to 9.2%, and 48.4% to 3.3% for PS, DPS, PSS_1, PSS_2, PFR, and PTS, respectively. The loss of hydroxyl groups due to the dehydration reaction during HTC results in lower the oxygen content [58]. The increase in carbon and a decrease in oxygen content complement the HHV increases after HTC treatment. Although the oxygen content decreases for DPS, the carbon content also decreases with HTC, resulting in the decrease of HHV. Hydrogen and nitrogen content change minimally and remain approximately the same. Fuel qualities of the PMS hydrochars were further analyzed with van-Krevelen diagram (Figure 1). In the van-Krevelen diagram, a sample closer to the origin indicates a higher quality fuel (i.e., anthracite) and a sample far from the origin specifies a poor-quality fuel (i.e., lignite or biomass). Fuel quality of all PMS except DPS being vastly upgraded with HTC at 260 °C and falls between bituminous and sub-bituminous region shown in Figure 1.

Ash content in each hydrochar was higher from its raw sample shown in Table 3. Ash is predominantly inert in the HTC process, so it is expected to become concentrated as the HTC process progresses. Higher ash content in treated samples is a result of HTC breaking down organic biomass constituents (release into liquid phase) and keep concentrating the inorganic content in the solid phase [25,61]. Similar to the ash content, lignin content was also higher in hydrochar compared to its raw sludge. This number increases from 5.2% to 36.9%, 10.0% to 19.7%, 20.2% to 46.8%, 17.6% to 87.8%, 22.6% to 57.9%, and 20.4% to 44.2% for PS, DPS, PSS_1, PSS_2, PFR, and PTS, respectively. As lignin breaks down between 255–265 °C [52,55], unreacted lignin is concentrated in the solid phase. Higher lignin content in hydrochar can be supported by the increase in HHV for most sludges following treatment as a result of reactive, less energy dense cellulose being removed from the char more at higher temperatures.

Total sugar content in raw samples and their hydrochar was almost the same, shown in Table 3. The short-chain sugar molecules (C_5 and C_6) are likely the product of hydrolyzed cellulose and hemicellulose [62]. However, these sugar molecules are highly reactive at HTC conditions and they tend to dehydrate into furan derivatives (e.g., furfural, hydroxymethyl furfural etc) or decarboxylated to short-chain acids, CO_2 and water [62]. However, in PMS, the number did not increase, which could indicate that dehydration and decarboxylation of sugars are more dominant than hydrolysis of PMS.

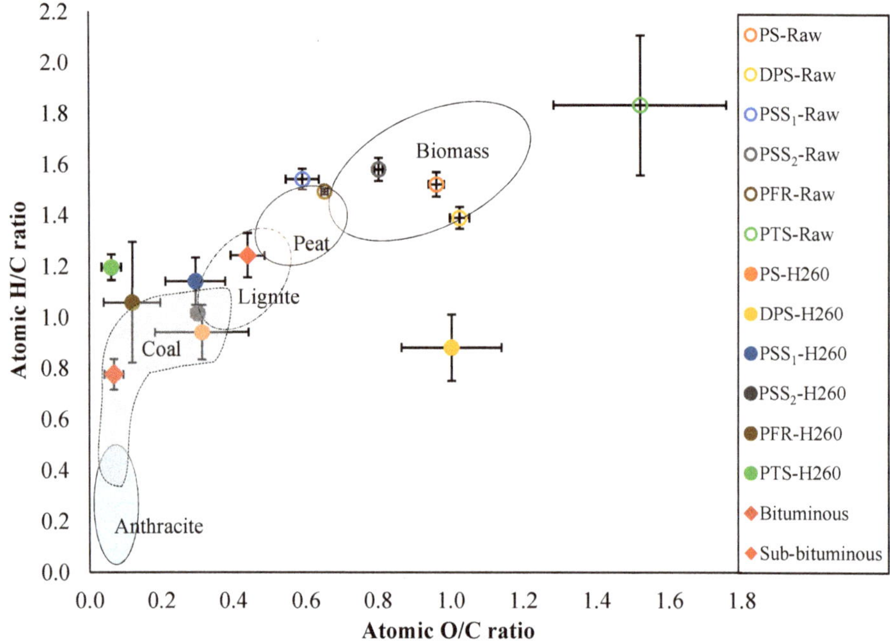

Figure 1. van-Krevelen diagram for all raw and hydrochar samples including bituminous and sub-bituminous coal.

Table 3. Ultimate analysis, ash and lignin number for all raw and hydrochar samples ([a] indicates that it was calculated by difference; [b] Below detection limit; and [c] Not applicable).

Sample	HTC Temperature (°C)	Carbon (%)	Hydrogen (%)	Nitrogen (%)	Sulfur (%)	Oxygen [a] (%)	Ash (%)	Lignin (%)	C_5 and C_6 Sugar (mg/g)
PS	Raw	34.6 ± 0.4	4.7 ± 0.1	0.3 ± 0.0	BD [b]	44.4 ± 0.8	16.0 ± 0.0	5.2 ± 0.4	2.7 ± 0.3
	260	44.3 ± 3.3	3.8 ± 0.3	0.2 ± 0.0	2.0 ± 0.0	18.5 ± 7.5	33.1 ± 1.7	36.9 ± 1.3	3.0 ± 0.3
DPS	Raw	27.1 ± 0.2	3.4 ± 0.1	0.7 ± 0.0	BD [b]	37.0 ± 0.9	31.9 ± 0.4	10.0 ± 1.2	2.0 ± 0.2
	260	22.2 ± 1.7	1.8 ± 0.2	0.4 ± 0.1	0.1 ± 0.0	29.6 ± 3.4	46.0 ± 0.3	19.7 ± 0.8	1.9 ± 0.3
PSS_1	Raw	38.4 ± 0.8	5.3 ± 0.1	2.3 ± 0.2	0.9 ± 0.1	30.3 ± 2.2	22.8 ± 0.4	20.2 ± 0.1	4.0 ± 1.1
	260	40.5 ± 1.5	4.2 ± 0.3	2.1 ± 0.0	1.3 ± 0.0	16.0 ± 4.4	36.0 ± 1.3	46.8 ± 0.8	1.4 ± 0.0
PSS_2	Raw	42.8 ± 0.0	6.1 ± 0.2	0.7 ± 0.1	0.2 ± 0.0	45.8 ± 0.7	4.5 ± 0.1	17.6 ± 0.5	3.5 ± 1.2
	260	62.2 ± 0.7	5.7 ± 0.1	1.6 ± 0.1	0.3 ± 0.0	25.0 ± 1.5	5.2 ± 0.1	87.8 ± 5.9	3.5 ± 2.5
PFR	Raw	39.8 ± 0.1	5.3 ± 0.0	0.6 ± 0.1	0.3 ± 0.0	34.6 ± 0.5	19.5 ± 0.1	22.6 ± 0.3	3.8 ± 2.2
	260	57.6 ± 2.6	5.5 ± 1.2	0.6 ± 0.1	0.4 ± 0.1	9.2 ± 6.1	26.8 ± 0.3	57.9 ± 1.2	2.8 ± 0.3
PTS	Raw	23.9 ± 2.7	3.9 ± 0.4	0.6 ± 0.0	6.2 ± 0.6	48.4 ± 5.2	17.1 ± 0.1	20.4 ± 0.5	3.2 ± 2.2
	260	40.2 ± 1.4	4.3 ± 0.1	0.4 ± 0.1	1.3 ± 0.2	3.3 ± 1.5	50.5 ± 1.5	44.2 ± 0.6	2.4 ± 1.2
Clarion # 5a ccal	Raw	63.8 ± 1.2	4.1 ± 0.1	1.5 ± 0.0	4.6 ± 0.8	$14.9 \pm$	11.1 ± 0.6	NA [c]	NA [c]
Bituminous coal [63]	Raw	75.1 ± 2.8	5.3 ± 0.4	1.5 ± 0.1	1.1 ± 0.8	6.9 ± 2.6	10.0 ± 1.4	NA [c]	NA [c]
Sub-bituminous ccal [64]	Raw	55.8 ± 3.5	6.2 ± 0.2	0.7 ± 0.1	0.3 ± 0.1	32.7 ± 2.8	4.3 ± 0.9	NA [c]	NA [c]

3.3. Fuel Characteristics of Hydrochars

Carbon contents in the PMS hydrochars were increased with HTC, while oxygen contents were decreased, resulting in increases of HHV_{daf}. While this increase makes the hydrochars more similar to bituminous coal, another factor to consider is that there is clearly a wide range of ash contents observed in PMS samples, which will impact fuel quality. Additionally, combustion properties of those hydrochars are required to further evaluate the fuel characteristics in order to determine the feasibility of co-combustion with coal. Pyrolysis thermograms of various hydrochars, shown in Figure 2, illustrates the thermal gravimetric (TG) curves and derivative thermal gravimetric (DTG) curves for hydrochars produced at 260 °C from PS, DPS, PSS_1, PSS_2, PFR, and PTS with respect to TG pyrolysis temperatures. The TG curves for PS-H260, PSS_2-H260, PFR-H260, and PTS-H260 showed a rapid mass loss between 190 to 550 °C with a maximum mass loss rate of −4.14, −7.86, 6.05, and −6.48% min^{-1}, respectively. This rapid mass loss could be attributed to the high volatile materials present in these hydrochars [65]. The PSS_2-H260 lost 61% of its initial mass during the complete pyrolysis stage which was the maximum among others while the PS-H260, PFR-H260 and PTS-H260 have lost 51%, 55%, and 43%, respectively. Although PSS_1 and PSS_2 both are mixtures of primary and secondary sludges, the TG curve of PSS_1-H260 exhibited a slow mass loss for the same temperature range with a maximum mass loss rate of −3.21% min^{-1} compared to PSS_2-H260 and lost 46% of its initial mass. This is another confirmation that the sources of these samples were different and/or the ratios of primary to secondary sludges are not the same. The TG curve of DPS-H260 displayed a maximum mass loss rate of −7.13% min^{-1} at a much higher temperature range of 640–780 °C. The presence of higher inorganic compounds or lignin could be the possible reason for this shift towards higher temperature compared to other sludges [66]. The TG curve for PS, which is dewatered PTS [14], also illustrated its maximum mass loss rate in this temperature range.

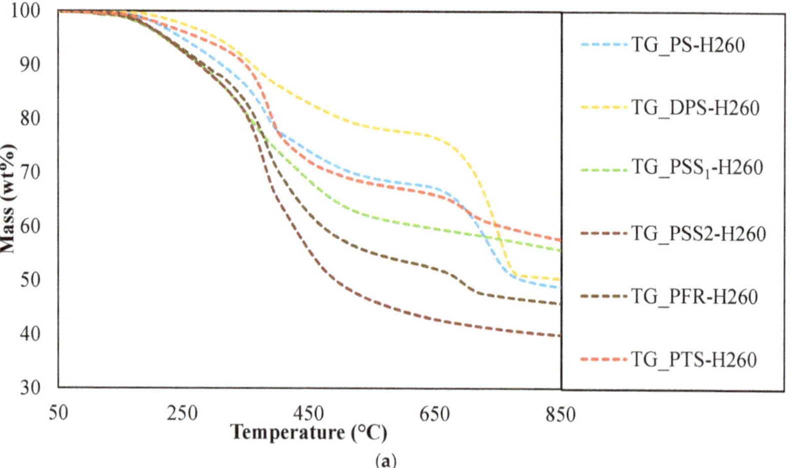

(a)

Figure 2. *Cont.*

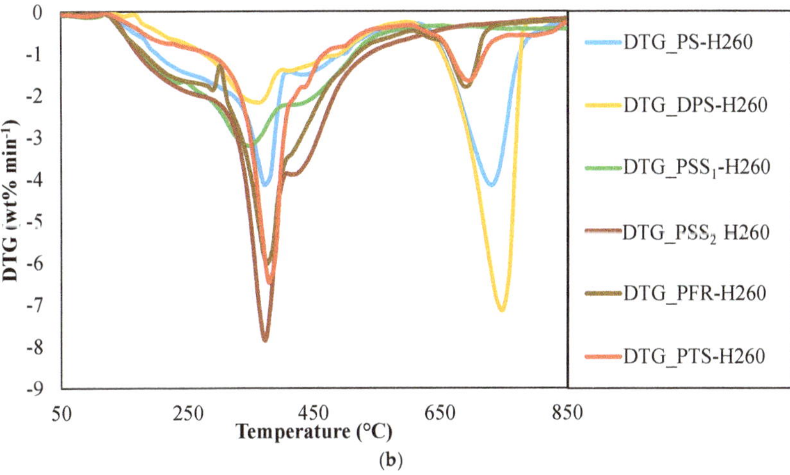

(b)

Figure 2. Pyrolysis thermograph plotted versus temperature (**a**) and derivative thermal gravimetric (DTG) curves plotted versus temperature (**b**) for all 260 °C hydrothermal carbonization (HTC) treated paper mill sludges (PMS).

The DTG curves for all the hydrochars, except for PSS_1-H260 and PSS_2-H260, showed two distinct peaks at two different temperature ranges of 310–400°C and 640–780°C, respectively. The DTG curves of PSS_1-H260 and PSS_2-H260 displayed only one major peak in between 310–400 °C. All DTG curves have minor peaks after the major peak in the temperature range of 310–400°C. The origin of these peaks can be explained from the constituents of PMS. Yanfen et al. [66] stated that paper pulp and fiber are the main organic elements of PMS and because of the degradation of hemicellulose, cellulose, and lignin present in the sludges are primarily responsible for these peaks. Previous studies have also found that hemicellulose decomposes between 200–300 °C during pyrolysis, while cellulose decomposes between 300–400 °C. The pyrolysis of lignin starts above 300 °C and continues for a wide temperature range [58]. All the peaks illustrated in DTG curve of Figure 2b agree with these findings. The peaks before 400 °C attributed to the complete decomposition of hemicellulose and cellulose and partial decomposition of lignin as it degrades over a wide range of pyrolysis temperature. The peaks after 400 °C correspond to the complete breaking of lignin [50]. However, the second peak for PS and DPS emerged at 750 °C. The pyrolysis of chemicals (i.e., additives, coagulants, flocculants, $CaCO_3$, and other minerals) present in the PS and DPS could be the possible reason for the emergence of the second peak at a higher temperature compared to other sludges [66,67].

The proximate analysis of various raw and PMS hydrochars at 180, 220, and 260 °C are presented in Figure 3. The figure indicates that the raw PMS samples were high in volatile matters and low in fixed carbon. Among the raw PMS samples, PSS_2 contained the highest volatile matters and the lowest ash while PSS_1 had the highest fixed carbon. Figure 3 also shows the change of ash, fixed carbon, and volatile matters present in various sludges with respect to HTC temperatures compared to raw samples. The figure also illustrates that HTC of PMS at 180 and 220 °C did not significantly affect the volatile materials, fixed carbon, and ash content compared to raw PMS samples, which have already been discussed in Section 3.1. However, the ash and fixed carbon were increased and volatile matters of PMS decreased remarkably when HTC temperature increased from 220 °C to 260 °C. The presence of high inorganic materials, which were trapped in the solid phase of the product instead of being leached in to the liquid phase, could be a possible reason for the increase in ash. The decrease in volatile materials could be attributed to the decomposition of cellulose and hemicellulose producing gaseous species like CO, CO_2, and short-chained hydrocarbons [66,68]. The decomposition of volatile materials also increased the fixed carbon in 260 °C hydrochars [69]. Among the PMS hydrochars, PSS_2 contained the maximum volatile materials and fixed carbon as well as the lowest ash percentage.

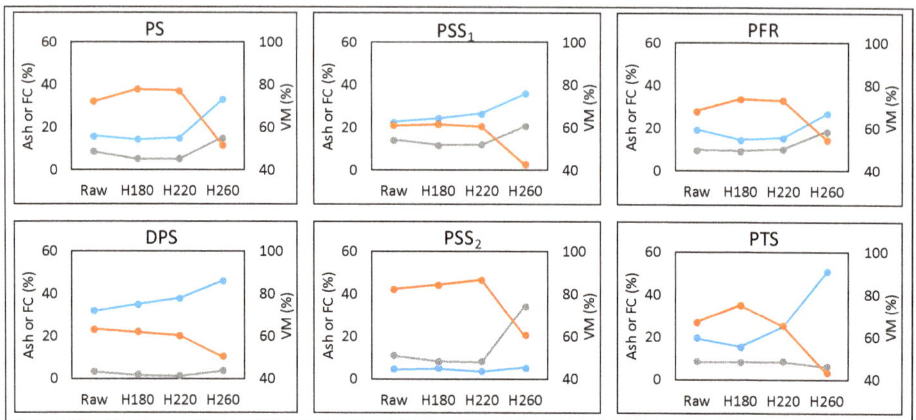

Figure 3. Proximate analysis of various paper mill sludges hydrothermally carbonized at 180–260 °C. The orange, gray, and blue lines represent volatile matter (VM), fixed carbon (FC) and ash content, respectively.

Figure 4 shows the co-combustion thermograph of coal, hydrochar of one paper mill sludge PTS, at 260 °C and their 50-50 mix. These parameters were selected because the PTS hydrochar saw the greatest energy densification at 260°C. Additionally, 50% mass should allow for both components to show combustion characteristics to determine if any positive or negative synergistic effects exist during co-firing. The PTS hydrothermally treated at 260 °C was co-combusted with coal to examine the combustion characteristics through TG curve and first negative derivative of DTG curve. The TG curve of this PTS hydrochar showed a rapid mass loss of approximately 60% from 170–350 °C. This rapid mass loss occurred due to the presence of high volatile materials (~45%) in PTS hydrochar [59]. However, the TG curve for coal displayed a comparatively slow mass loss from 170–608°C as it contained lower volatile materials than PTS hydrochar treated at 260 °C. The combustion TG curve for the mix followed a trend similar to the coal TG curve as it showed a slow mass loss in the same temperature range as coal. The DTG curve for PTS hydrochar indicated a maximum mass loss rate of −35.71% min^{-1} at 317 °C whereas coal DTG curve showed the maximum mass loss rate at a higher temperature of 537 °C. The mix DTG showed a combustion characteristic more like the coal where the maximum mass loss occurred at 541 °C. It is observed from Figure 4 that the combustion of 50-50 mix of PTS hydrochar and coal was more similar to the coal combustion than to PTS combustion.

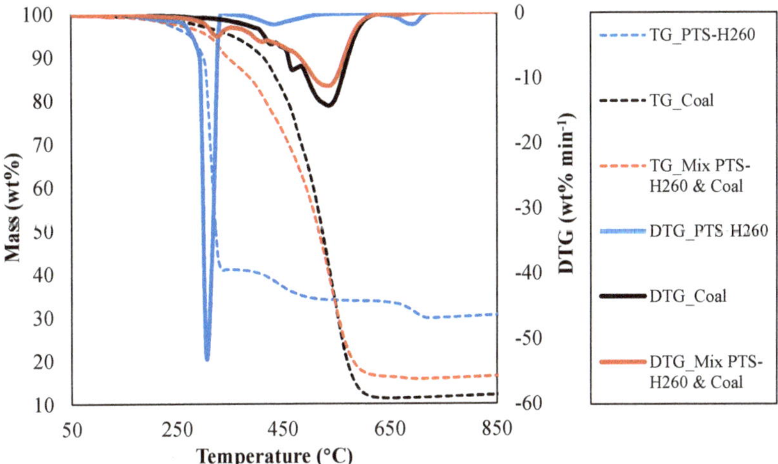

Figure 4. Combustion thermograph and DTG curves plotted versus temperature for 260 °C HTC treated paper mill sludges, coal and 50% mixtures of coal and hydrochar (TG nomenclature refers to mass loss on the left axis and DTG can be found on the right axis).

Additional efforts were made to analyze the combustion characteristics of Clarion # 5a coal (volatile matter (VM) and fixed carbon (FC) are 34.3 $\pm$ 0.0 and 54.0 $\pm$ 0.5, respectively), all the HTC-260 hydrochars and their 50-50 (wt% dry basis) mix with coal. During co-combustion, DPS-H260 was not considered as it has low ED compared to other PMS as discussed in earlier sections. All the parameters of combustion characteristics are presented in Table 4. Ignition and burnout indices were calculated using Equations (6) and (7). The ignition temperatures for all the hydrochars produced at 260 °C, except for DPS-H260, were low compared to coal as they contain high volatile material than coal [70]. The presence of high inorganics in DPS could be a reason for high ignition temperature. The other combustion parameters of the same hydrochar also deviated a lot compared to coal. However, when other 260's hydrochars were mixed with coal, these deviations became small. For example, before mixing, the values of ignition temperature, ignition time, burnout temperature, burnout time, DTG_{max}, DTG_{max} time, ignition and burnout indices for PTS-H260 were 294.2 °C, 20.3 min, 342.8 °C, 22.8 min, -52.1% min^{-1}, 20.8 min, 1.24×10^{-1}, and 5.40×10^{-3}, respectively. When mixed with coal, these values became 430.5 °C, 26.8 min, 692.0°C, 39.8 min, -11.1% min^{-1}, 31.8 min, 1.31×10^{-2}, and 3.12×10^{-4}, respectively, which were close to the values for coal. However, in the case of PS-H260, the deviations were still large even though the DTG_{max} value became closer than all the other hydrochars. As such, it is evident from Table 4 that combusting hydrochar produced from PMS at 260 °C (except PS) with coal made had a co-combusting characteristic that is similar to coal.

Table 4. Combustion Characteristics of determined from Combustion TGA data. (Clarion #5a coal (bituminous) was used for co-combustion [59]).

Sample	Ignition Temperature (°C)	Ignition Time (min)	DTG_{max} (% min^{-1})	Burnout Temperature (°C)	Burnout Time (min)	DTG_{max} Time (min)	Ignition Index, D_i	Burnout Index, D_B
PS-H260	380.4	23.8	−19.4	752.5	42.3	25.8	3.17×10^{-2}	7.19×10^{-4}
DPS-H260	628.1	36.3	−6.6	788.7	44.3	41.8	4.39×10^{-3}	9.10×10^{-5}
PSS_1-H260	262.0	18.8	−8.6	606.4	35.8	20.8	2.21×10^{-2}	5.88×10^{-4}
PSS_2-H260	273.0	19.3	−41.7	557.1	33.3	19.8	1.10×10^{-1}	3.26×10^{-3}
PFR-H260	294.1	20.3	−38.0	708.4	40.8	21.3	8.82×10^{-2}	2.10×10^{-3}
PTS-H260	294.2	20.3	−52.1	718.5	41.3	20.8	1.24×10^{-1}	2.98×10^{-3}
Coal	466.7	28.8	−14.2	627.8	36.8	32.3	1.53×10^{-2}	4.19×10^{-4}
50% Coal + 50% PS-H260	305.3	20.8	−14.5	739.4	42.3	21.8	3.22×10^{-2}	7.49×10^{-4}
50% Coal + 50% PSS_1-H260	386.3	24.8	−7.0	648.0	37.8	31.8	8.97×10^{-3}	2.10×10^{-4}
50% Coal + 50% PSS_2-H260	398.7	24.8	−9.9	660.3	37.8	31.3	1.28×10^{-2}	3.16×10^{-4}
50% Coal + 50% PFR-H260	410.3	25.3	−7.5	701.8	39.8	31.8	9.38×10^{-3}	2.46×10^{-4}
50% Coal + 50% PTS-H260	430.5	26.8	−11.1	692.0	39.8	31.8	1.31×10^{-2}	3.12×10^{-4}

4. Conclusions

Six different paper mill sludges have been hydrothermally carbonized at 180 °C, 220 °C, and 260 °C. With the exception of DPS; HHV_{daf} and ED ratio were maximized at HTC temperature of 260 °C for all the PMS. Although PMS sludges showed two-staged thermal degradation during pyrolysis, the thermograms revealed that hydrochars produced from different PMS possess different degradation severity. While, primary and secondary mix sludges show lower solid mass yields at lower HTC temperatures, the deinked primary sludge shows a higher solid mass yield at similar HTC conditions. Moreover, primary and secondary sludge samples showed anomalies in thermograms between themselves. The combustion indices indicate that hydrochar produced from paper mill sludges have lower ignition temperatures and higher burnout temperatures than raw bituminous coal. However, co-firing with coal and hydrochar showed higher ignition temperatures and lower burnout temperatures. Moreover, most of the hydrochars, specifically hydrochars from primary and secondary sludge, have shown very similar combustion thermograms to coal. In addition, by considering the ash content in the feedstocks and their hydrochars, PSS_2 could be a suitable co-fired option in existing coal-fired power plants.

Author Contributions: N.S., A.S., and K.M. performed the HTC experiments, analyzed the results and wrote the article draft. P.S. performed TGA experiments, analyzed the results, and wrote the corresponding section. D.F.-V. and W.M.H.-C. performed sugar and lignin analyses and analyzed the results. W.J.O. and M.T.R. reviewed the article.

Funding: This research is funded by Ohio Coal Development Office, grant number R-17-05 and Ohio University start-up funding support.

Acknowledgments: The authors acknowledge Wen Fan and Ravi Garlapalli for their supports in the laboratory. The authors are also thankful to Rifat Hasan and Alex Flamm at the Institute for Sustainable Energy and the Environment (ISEE) for their laboratory efforts in this project.

Conflicts of Interest: The authors declare no conflicts of interest.

References

1. Shaykheeva, D.; Panasyuk, M.; Malganova, I.; Khairullin, I. World Population Estimates and Projections: Data and Methods. *J. Econ. Econ. Educ. Res.* **2016**, *17*, 237–247.
2. Annual Energy Outlook 2016. Available online: http://large.stanford.edu/courses/2016/ph240/martelaro1/docs/0383-2016.pdf (accessed on 5 March 2019).
3. United States Department of Energy. *Billion Ton Update: Biomass Supply for a Bioenergy and Bioproducts Industry*; Oak Ridge National Laboratory: Oak Ridge, TN, USA, 2016.
4. Bird, M.; Talberth, J. *Waste Stream Reduction and Re-Use in the Pulp and Paper Sector*; Washington State Department of Ecology: Santa Fe, NM, USA, 2008.
5. Aspitarte, T.R.; Rosenfield, A.S.; Smale, B.C.; Amberg, H.R. *Methods for Pulp and Paper Mill Sludge Utilization and Disposal*; USEPA: Washington, DC, USA, 1973.
6. Mun, T.Y.; Tumsa, T.Z.; Lee, U.; Yang, W. Performance evaluation of co-firing various kinds of biomass with low rank coals in a 500 MWe coal-fired power plant. *Energy* **2016**, *115*, 954–962. [CrossRef]
7. Tsai, M.-Y.; Wu, K.-T.; Huang, C.-C.; Lee, H.-T. Co-firing of paper mill sludge and coal in an industrial circulating fluidized bed boiler. *Waste Manag.* **2002**, *22*, 439–442. [CrossRef]
8. Bajpai, P. Generation of Waste in Pulp and Paper Mills. In *Management of Pulp and Paper Mill Waste*; Springer International Publishing: Basel, Switzerland, 2015.
9. Monte, M.C.; Fuente, E.; Blanco, A.; Negro, C. Waste management from pulp and paper production in the European Union. *Waste Manag.* **2009**, *29*, 293–308. [CrossRef] [PubMed]
10. IPPC. *Reference Document on Best Available Techniques in the Pulp and Paper Industry*; European IPPC Bureau: Seville, Spain, 2001.
11. Soucy, J.; Koubaa, A.; Migneault, S.; Riedl, B. Chemical Composition and Surface Properties of Paper Mill Sludge and their Impact on High Density Polyethylene (HDPE) Composites. *J. Wood Chem. Technol.* **2016**, *36*, 77–93. [CrossRef]
12. Bajpai, P. *Recycling and Deinking of Recovered Paper*; Elsevier Science: Amsterdam, The Netherlands, 2013.

13. CANMET. *Pulp and Paper Sludge to Energy—Preliminary Assessment of Technologies, Canada*; CANMET Energy Technology Centre: Ottawa, ON, Canada, 2005.
14. Scott, G.M.; Smith, A. Sludge characteristics and disposal alternatives for the pulp and paper industry. In Proceedings of the 1995 International Environmental Conference, Atlanta, GA, USA, 7–10 May 1995; pp. 269–279.
15. Reza, M.T. *Upgrading Biomass by Hydrothermal and Chemical Conditioning*; University of Nevada Reno: Reno, NV, USA, 2013.
16. Reza, M.T.; Andert, J.; Wirth, B.; Busch, D.; Pielert, J.; Lynam, J.G.; Mumme, J. Hydrothermal Carbonization of Biomass for Energy and Crop Production. *Appl. Bioenergy* **2014**, *1*, 11–29. [CrossRef]
17. Reza, M.T.; Coronella, C.; Holtman, K.M.; Franqui-Villanueva, D.; Poulson, S.R. Hydrothermal Carbonization of Autoclaved Municipal Solid Waste Pulp and Anaerobically Treated Pulp Digestate. *ACS Sustain Chem. Eng.* **2016**, *4*, 3649–3658. [CrossRef]
18. Reza, M.T.; Mumme, J.; Ebert, A. Characterization of Hydrochar Obtained from Hydrothermal Carbonization of Wheat Straw Digestate. *Biomass Convers. Biorefin.* **2015**. [CrossRef]
19. Kruse, A.; Dinjus, E. Hot compressed water as reaction medium and reactant: Properties and synthesis reactions. *J. Supercrit. Fluids* **2007**, *39*, 362–380. [CrossRef]
20. Bandura, A.V.; Lvov, S.N. The ionization constant of water over wide ranges of temperature and density. *J. Phys. Chem. Ref. Data* **2006**, *35*, 15–30. [CrossRef]
21. Reza, M.T.; Uddin, M.H.; Lynam, J.; Hoekman, S.K.; Coronella, C. Hydrothermal carbonization of loblolly pine: Reaction chemistry and water balance. *Biomass Convers. Biorefin.* **2014**, *4*, 311–321. [CrossRef]
22. Reza, M.T.; Rottler, E.; Herklotz, L.; Wirth, B. Hydrothermal carbonization (HTC) of wheat straw: Influence of feedwater pH prepared by acetic acid and potassium hydroxide. *Bioresour. Technol.* **2015**, *182*, 336–344. [CrossRef] [PubMed]
23. Reza, M.T.; Wirth, B.; Luder, U.; Werner, M. Behavior of selected hydrolyzed and dehydrated products during hydrothermal carbonization of biomass. *Bioresour. Technol.* **2014**, *169*, 352–361. [CrossRef] [PubMed]
24. Reza, M.T.; Yan, W.; Uddin, M.H.; Lynam, J.G.; Hoekman, S.K.; Coronella, C.J.; Vasquez, V.R. Reaction kinetics of hydrothermal carbonization of loblolly pine. *Bioresour. Technol.* **2013**, *139*, 161–169. [CrossRef] [PubMed]
25. Lynam, J.G.; Reza, M.T.; Yan, W.; Vásquez, V.R.; Coronella, C.J. Hydrothermal carbonization of various lignocellulosic biomass. *Biomass Convers. Biorefin.* **2015**, *5*, 173–181. [CrossRef]
26. Makela, M.; Benavente, V.; Fullana, A. Hydrothermal carbonization of lignocellulosic biomass: Effect of process conditions on hydrochar properties. *Appl. Energy* **2015**, *155*, 576–584. [CrossRef]
27. Reza, M.T.; Lynam, J.G.; Vasquez, V.R.; Coronella, C.J. Pelletization of Biochar from Hydrothermally Carbonized Wood. *Environ. Prog. Sustain. Energy* **2012**, *31*, 225–234. [CrossRef]
28. Reza, M.T.; Yang, X.; Coronella, C.J.; Lin, H.; Hathwaik, U.; Shintani, D.; Neupane, B.P.; Miller, G.C. Hydrothermal Carbonization (HTC) and Pelletization of Two Arid Land Plants Bagasse for Energy Densification. *ACS Sustain. Chem. Eng.* **2015**. [CrossRef]
29. Saha, N.; Saba, A.; Reza, M.T. Effect of hydrothermal carbonization temperature on pH, dissociation constants, and acidic functional groups on hydrochar from cellulose and wood. *J. Anal. Appl. Pyrolysis* **2019**, *137*, 138–145. [CrossRef]
30. Naisse, C.; Alexis, M.; Plante, A.F.; Wiedner, K.; Glaser, B.; Pozzi, A.; Carcaillet, C.; Criscuoli, I.; Rumpel, C. Can biochar and hydrochar stability be assessed with chemical methods? *Org. Geochem.* **2013**, *60*, 40–44. [CrossRef]
31. Wiedner, K.; Naisse, C.; Rumpel, C.; Pozzi, A.; Wieczorek, P.; Glaser, B. Chemical modification of biomass residues during hydrothermal carbonization—What makes the difference, temperature or feedstock? *Org. Geochem.* **2013**, *54*, 91–100. [CrossRef]
32. Wiedner, K.; Rumpel, C.; Steiner, C.; Pozzi, A.; Maas, R.; Glaser, B. Chemical evaluation of chars produced by thermochemical conversion (gasification, pyrolysis and hydrothermal carbonization) of agro-industrial biomass on a commercial scale. *Biomass Bioenergy* **2013**, *59*, 264–278. [CrossRef]
33. Ronix, A.; Pezoti, O.; Souza, L.S.; Souza, I.P.A.F.; Bedin, K.C.; Souza, P.S.C.; Silva, T.L.; Melo, S.A.R.; Cazetta, A.L.; Almeida, V.C. Hydrothermal carbonization of coffee husk: Optimization of experimental parameters and adsorption of methylene blue dye. *J. Environ. Chem. Eng.* **2017**, *5*, 4841–4849. [CrossRef]

34. Kannan, S.; Gariepy, Y.; Vijaya Raghavan, G.S. Optimization of the conventional hydrothermal carbonization to produce hydrochar from fish waste. *Biomass Convers. Biorefin.* **2018**, *8*, 563–576. [CrossRef]
35. Kruse, A.; Zevaco, A.T. Properties of Hydrochar as Function of Feedstock, Reaction Conditions and Post-Treatment. *Energies* **2018**, *11*, 674. [CrossRef]
36. Román, S.; Libra, J.; Berge, N.; Sabio, E.; Ro, K.; Li, L.; Ledesma, B.; Álvarez, A.; Bae, S. Hydrothermal Carbonization: Modeling, Final Properties Design and Applications: A Review. *Energies* **2018**, *11*, 216. [CrossRef]
37. Kempegowda, R.S.; Skreiberg, Ø.; Tran, K.-Q.; Selvam, P.V.P. Techno-economic Assessment of Thermal Co-pretreatment and Co-digestion of Food Wastes and Sewage Sludge for Heat, Power and Biochar Production. *Energy Procedia* **2017**, *105*, 1737–1742. [CrossRef]
38. Escala, M.; Zumbuhl, T.; Koller, C.; Junge, R.; Krebs, R. Hydrothermal Carbonization as an Energy-Efficient Alternative to Established Drying Technologies for Sewage Sludge: A Feasibility Study on a Laboratory Scale. *Energy Fuel* **2013**, *27*, 454–460. [CrossRef]
39. Lucian, M.; Fiori, L. Hydrothermal Carbonization of Waste Biomass: Process Design, Modeling, Energy Efficiency and Cost Analysis. *Energies* **2017**, *10*, 211. [CrossRef]
40. Areeprasert, C.; Coppola, A.; Urciuolo, M.; Chirone, R.; Yoshikawa, K.; Scala, F. The effect of hydrothermal treatment on attrition during the fluidized bed combustion of paper sludge. *Fuel Process. Technol.* **2015**, *140*, 57–66. [CrossRef]
41. Areeprasert, C.; Zhao, P.T.; Ma, D.C.; Shen, Y.F.; Yoshikawa, K. Alternative Solid Fuel Production from Paper Sludge Employing Hydrothermal Treatment. *Energy Fuel* **2014**, *28*, 1198–1206. [CrossRef]
42. Zhao, P.T.; Ge, S.F.; Ma, D.C.; Areeprasert, C.; Yoshikawa, K. Effect of Hydrothermal Pretreatment on Convective Drying Characteristics of Paper Sludge. *ACS Sustain. Chem. Eng.* **2014**, *2*, 665–671. [CrossRef]
43. Areeprasert, C.; Ma, D.; Prayoga, P.; Yoshikawa, K. A Review on Pilot-Scale Applications of Hydrothermal Treatment for Upgrading Waste Materials. *Int. J. Environ. Sci. Dev.* **2016**, *7*, 425–430. [CrossRef]
44. Areeprasert, C.; Scala, F.; Coppola, A.; Urciuolo, M.; Chirone, R.; Chanyavanich, P.; Yoshikawa, K. Fluidized bed co-combustion of hydrothermally treated paper sludge with two coals of different rank. *Fuel Process Technol.* **2016**, *144*, 230–238. [CrossRef]
45. Mäkelä, M.; Forsberg, J.; Söderberg, C.; Larsson, S.H.; Dahl, O. Process water properties from hydrothermal carbonization of chemical sludge from a pulp and board mill. *Bioresour. Technol.* **2018**, *263*, 654–659. [CrossRef] [PubMed]
46. Makela, M.; Benavente, V.; Fullana, A. Hydrothermal carbonization of industrial mixed sludge from a pulp and paper mill. *Bioresour. Technol.* **2016**, *200*, 444–450. [CrossRef] [PubMed]
47. TAPPI. *Acid-Insoluble Lignin in Wood and Pulp*; (Reaffirmation of T 222 om-02); TAPPI: Peachtree Corners, GA, USA, 2006; Available online: https://www.tappi.org/content/SARG/T222.pdf (accessed on 4 March 2019).
48. Sequeira, R.M.; Lew, R.B. Carbohydrate composition of almond hulls. *J. Agric. Food Chem.* **1970**, *18*, 950–951. [PubMed]
49. Rong, H.; Wang, T.; Zhou, M.; Wang, H.; Hou, H.; Xue, Y. Combustion Characteristics and Slagging during Co-Combustion of Rice Husk and Sewage Sludge Blends. *Energies* **2017**, *10*, 438. [CrossRef]
50. Vamvuka, D.; El Chatib, N.; Sfakiotakis, S. Measurements of ignition point and combustion characteristics of biomass fuels and their blends with lignite. *Combustion* **2011**, *2015*, 95.
51. Funke, A.; Ziegler, F. Hydrothermal carbonization of biomass: A summary and discussion of chemical mechanisms for process engineering. *Biofuels Bioprod. Biorefin.* **2010**, *4*, 160–177. [CrossRef]
52. Hoekman, S.K.; Broch, A.; Robbins, C. Hydrothermal Carbonization (HTC) of Lignocellulosic Biomass. *Energy Fuel* **2011**, *25*, 1802–1810. [CrossRef]
53. Sheng, C.; Azevedo, J.L.T. Estimating the higher heating value of biomass fuels from basic analysis data. *Biomass Bioenergy* **2005**, *28*, 499–507. [CrossRef]
54. Zhao, C.; Jiang, E.; Chen, A. Volatile production from pyrolysis of cellulose, hemicellulose and lignin. *J. Energy Inst.* **2017**, *90*, 902–913. [CrossRef]
55. Kang, S.; Li, X.; Fan, J.; Chang, J. Characterization of Hydrochars Produced by Hydrothermal Carbonization of Lignin, Cellulose, d-Xylose, and Wood Meal. *Ind. Eng. Chem. Res.* **2012**, *51*, 9023–9031. [CrossRef]
56. Lin, Y.; Ma, X.; Peng, X.; Hu, S.; Yu, Z.; Fang, S. Effect of hydrothermal carbonization temperature on combustion behavior of hydrochar fuel from paper sludge. *Appl. Therm. Eng.* **2015**, *91*, 574–582. [CrossRef]

57. Kim, J.S.; Lee, Y.Y.; Park, S.C. Pretreatment of Wastepaper and Pulp Mill Sludge by Aqueous Ammonia and Hydrogen Peroxide. In *Twenty-First Symposium on Biotechnology for Fuels and Chemicals, Proceedings of the Twenty-First Symposium on Biotechnology for Fuels and Chemicals, Fort Collins, CO, USA, 2–6 May 1999*; Finkelstein, M., Davison, B.H., Eds.; Humana Press: Totowa, NJ, USA, 2000; pp. 129–139. [CrossRef]
58. Saba, A.; Saha, P.; Reza, M.T. Co-Hydrothermal Carbonization of coal-biomass blend: Influence of temperature on solid fuel properties. *Fuel Process. Technol.* **2017**, *167*, 711–720. [CrossRef]
59. García, G.; Arauzo, J.; Gonzalo, A.; Sánchez, J.L.; Ábrego, J. Influence of feedstock composition in fluidised bed co-gasification of mixtures of lignite, bituminous coal and sewage sludge. *Chem. Eng. J.* **2013**, *222*, 345–352. [CrossRef]
60. Grénman, H.; Eränen, K.; Krogell, J.; Willför, S.; Salmi, T.; Murzin, D.Y. Kinetics of Aqueous Extraction of Hemicelluloses from Spruce in an Intensified Reactor System. *Ind. Eng. Chem. Res.* **2011**, *50*, 3818–3828. [CrossRef]
61. McGaughy, K.; Reza, M.T. Recovery of Macro and Micro-Nutrients by Hydrothermal Carbonization of Septage. *J. Agric. Food Chem.* **2018**, *66*, 1854–1862. [CrossRef] [PubMed]
62. Wang, T.; Zhai, Y.; Zhu, Y.; Li, C.; Zeng, G. A review of the hydrothermal carbonization of biomass waste for hydrochar formation: Process conditions, fundamentals, and physicochemical properties. *Renew. Sustain. Energy Rev.* **2018**, *90*, 223–247. [CrossRef]
63. McKendry, P. Energy production from biomass (part 1): Overview of biomass. *Bioresour. Technol.* **2002**, *83*, 37–46. [CrossRef]
64. Stricker, G.D.; Flores, R.M.; Trippi, M.H.; Ellis, M.S.; Olson, C.M.; Sullivan, J.E.; Takahashi, K.I. *Coal Quality and Major, Minor, and Trace Elements in the Powder River, Green River, and Williston basins, Wyoming and North Dakota: U.S. Geological Survey*; U.S. Department of the Interior: Reston, VA, USA, 2007.
65. El-Sayed, S.A.; Mostafa, M. Pyrolysis characteristics and kinetic parameters determination of biomass fuel powders by differential thermal gravimetric analysis (TGA/DTG). *Energy Convers. Manag.* **2014**, *85*, 165–172. [CrossRef]
66. Yanfen, L.; Xiaoqian, M. Thermogravimetric analysis of the co-combustion of coal and paper mill sludge. *Appl. Energy* **2010**, *87*, 3526–3532. [CrossRef]
67. Volpe, M.; Goldfarb, J.L.; Fiori, L. Hydrothermal carbonization of Opuntia ficus-indica cladodes: Role of process parameters on hydrochar properties. *Bioresour. Technol.* **2018**, *247*, 310–318. [CrossRef] [PubMed]
68. Gao, Y.; Wang, X.; Wang, J.; Li, X.; Cheng, J.; Yang, H.; Chen, H. Effect of residence time on chemical and structural properties of hydrochar obtained by hydrothermal carbonization of water hyacinth. *Energy* **2013**, *58*, 376–383. [CrossRef]
69. He, C.; Giannis, A.; Wang, J.-Y. Conversion of sewage sludge to clean solid fuel using hydrothermal carbonization: Hydrochar fuel characteristics and combustion behavior. *Appl. Energy* **2013**, *111*, 257–266. [CrossRef]
70. Varol, M.; Atimtay, A.; Bay, B.; Olgun, H. Investigation of co-combustion characteristics of low quality lignite coals and biomass with thermogravimetric analysis. *Thermochim. Acta* **2010**, *510*, 195–201. [CrossRef]

Article

Techno-Economic Assessment of Co-Hydrothermal Carbonization of a Coal-Miscanthus Blend

Akbar Saba [1,2], Kyle McGaughy [1,3] and M. Toufiq Reza [1,2,*]

1 Institute for Sustainable Energy and the Environment, 1 Ohio University, Athens, OH 45701, USA; as030616@ohio.edu (A.S.); km067812@ohio.edu (K.M.)
2 Department of Mechanical Engineering, 1 Ohio University, Athens, OH 45701, USA
3 Department of Chemical and Biomolecular Engineering, 1 Ohio University, Athens, OH 45701, USA
* Correspondence: reza@ohio.edu; Tel.: +1-740-593-1506

Received: 4 January 2019; Accepted: 5 February 2019; Published: 15 February 2019

Abstract: Co-Hydrothermal Carbonization (Co-HTC) is a thermochemical process, where coal and biomass were treated simultaneously in subcritical water, resulting in bulk-homogenous hydrochar that is carbon-rich and a hydrophobic solid fuel with combustion characteristics like coal. In this study, technoeconomic analysis of Co-HTC was performed for a scaled-up Co-HTC plant that produces fuel for 110 MWe coal-fired power plant using Clarion coal #4a and miscanthus as starting feedstocks. With precise mass and energy balance of the Co-HTC process, sizing of individual equipment was conducted based on various systems equations. Cost of electricity was calculated from estimated capital, manufacturing, and operating and maintenance costs. The breakeven selling price of Co-HTC hydrochar was $117 per ton for a 110 MWe. Sensitivity analysis indicates that this breakeven selling price could be as low as $106 per ton for a higher capacity plant. Besides plant size, the price of solid fuel is sensitive to the feedstock costs and hydrochar yield.

Keywords: coal; biomass; hydrochar; process economics; sensitivity analysis; cost of electricity

1. Introduction

Coal dependence has led to significant air pollution as the combustion of coal releases various air pollutants (trace metals, chlorine, greenhouse gases etc.) that are hazardous and toxic to the surrounding life and environment. In 2016, U.S. coal combustion accounted for 26.3% of the CO_2 produced from fossil fuel combustion and contributed to nearly 69% (CO_2 equivalent basis) of the greenhouse gases produced from the electrical power generation sector [1]. In the same year, the electrical power sector also contributed to 43.8% of the total SO_2 released, of which 92% was contributed by coal combustion [1]. The mitigation of these pollutants while meeting the electricity demand are essential for keeping up with the continual growth of the world's economy as well as maintaining its environmental health. In the U.S., nearly 261 billion tons of coal are available in recoverable coal reserves (as of 2012) which is estimated to last for more than 150 years based on an average production rate of 1.5 billion tons per year [2]. Such resource availability allows for U.S. coal-fired power plants account for approximately 40% of yearly electricity generation (per 2015) while 2040 projections expect total consumption to be 705 million short tons (an approximate energy equivalent of 13.49 quadrillion) [3].

In 2017, biomass was the highest primary energy producer among renewable energies at 45.6% while accounting for 5.8% of the U.S. total primary energy produced (fossil fuels, nuclear, and renewables) [4,5]. Biomass utilization as a fuel can minimize greenhouse gas emission as the emitted carbon dioxide was originally used for plant growth via photosynthesis [4]. Thus, in an attempt to reduce coal-fired plant emissions in the form of toxic metals, sulfur dioxide, and nitrogen oxides, co-firing of biomass and coal has been implemented [6,7]. Despite the reduced environmental impacts from co-firing, significant combustion setbacks are still experienced. Overall energy content is reduced due to co-combustion of higher heating value of coal (e.g., 24–30 MJ kg^{-1}) with lower heating value of biomass (10–22 MJ kg^{-1}) [8–13]. Additionally, coal densities (e.g., 700–900 kg m^{-3} for bituminous coals) vary with biomass densities (e.g., 100–600 kg m^{-3}) which can result in density partioning and overall lack of homogeneity in the coal-biomass mixture [8–13]. Lastly, biomass processing (e.g., milling, drying, etc.) prior to co-firing are necessary installments that can be costly due high moisture content. In fact, a 2004 EERE study discussed that minor boiler modifications could be performed for existing coal plants in order to co-combust with biomass, however, feedstock handling for drying, grinding, and overall homogeneity would have to be maintained [14].

Hydrothermal carbonization (HTC) has been extensively researched for biomass pretreatment and low rank coal dewatering [15–19]. HTC uses subcritical water (temperatures 180–260 °C) to convert biomass into a carbon-rich and hydrophobic solid fuel, known as hydrochar. Co-Hydrothermal Carbonization (Co-HTC) is the simultaneous treatment of two feedstocks in the presence of subcritical water. Co-HTC can promote the synergistic interaction between the two feedstocks during treatment, which would not occur when the feedstocks are separately treated (i.e., HTC). Shen et al., demonstrated de-chlorination could be improved with Co-HTC [20]; as biomass intermediates produced during treatment provided more phenolic compounds and short chain organic intermediates to react with chlorine groups in PVC. Zhang et al., showed increased organic and carbon retention rates for the Co-HTC of sewage sludge and pine sawdust as opposed to the individual HTC of each respective feedstock [21]. It was hypothesized that Mailard reactions between sawdust sugar derivatives and sewage sludge protein formed insoluble hydrochar. More applicable to improved co-firing, Nonaka et al., and Saba et al., evaluated the Co-HTC of biomass-coal blends with varying feedstock compositions and varying reaction temperature, respectively. Nonaka et al., found that Co-HTC favored more polymerization reactions and that more thermally stable hydrochars were produced from higher coal ratios [22]. Furthermore, heating value and chemical characterization did not change notably with changing feedstock compositions. Meanwhile, Saba et al., observed hydrochar homogeneity and additional biomass degradation (via mass yield) catalyzed by Co-HTC with coal, resulting in lower pH media [23]; low pH would then encourage more leaching of inorganic content and sulfur. As expected, both studies found increased HHV of produced hydrochars than that of the blended and untreated feedstocks. Overall, hydrochar can potentially serve as a better option for co-firing than biomass; co-treatment with coal can have significant advantages as well.

HTC energetics have shown to be promising as the heat of reaction was shown to be exothermic by Funke and Ziegler [24]. For HTC of glucose, cellulose, and wood at 240 °C and 6 h heats of reactions were determined via digital scanning calorimetry to be approximately −1 MJ kg^{-1} for each feedstock. Increasing reaction severity through temperature (>310 °C) and time (64 h) resulted in a minimum heat of reaction of −2.4 MJ kg^{-1} as a result of the production of lower energy products CO_2 and H_2O [24]. Although consideration for reaction by-products were considered, all carbon losses from experimental mass balance was assumed to have gone into the gas phase as CO_2. These results show potential for alleviating energy input for a continuous process. In fact, utilization of these results allowed McGaughy et al., to perform an energy audit on a simplified continuous HTC plant that treats food waste [25]. Energy output to input ratio (EOIR) were 2.95–4.91 depending on HTC reaction temperature.

Although HTC treatment shows promising fuel upgrading and energy savings from decreasing energy duty for the feedstock drying step, process economics should also be considered for feasibility. Li et al., performed a techno-economic models for rice husk to fuel conversions via HTC, pyrolysis, and anaerobic digestion and compared them to the direct combustion of rice husks for heat and power [26]. With considering operating costs and negligible differences in fixed capital costs, HTC of rice husks proved to be a more advantageous process compared to the other two as it can have higher solids loading, lower utilities costs, and lower process water costs. For 1-ton rice husk conversion, HTC cost per MJ was \$0.013 MJ^{-1} (\$153 MWh^{-1}), having 81% and 38% savings compared to fossil fuel oil costs and direct combustion of rice husks (when not incorporating capital costs into the HTC costs), respectively. This cost is still 4.5 times more than the cost of Central Appalachian Coal at

\$33.7 MWh^{-1} [27]. Further heat integration of HTC process liquid can further benefit the process economics in this study with more temperature optimization of HTC. Additionally, considering capital costs, maintenance costs, and the time value of money contributes to a more accurate model for commercial design.

Lucian et al., did process modeling for the HTC of grape marc and off-specification compost from raw processing to pelletized hydrochar [28]. Of the various process parameters, process optimization from their data occurred at the shortest residence time of 1 h (from 1, 3, and 8 h times), a reaction temperature of 220 °C, and a slightly higher dry biomass-to-water treatment ratio of 0.19. Treatment at 220 °C from 180, 220, and 250 °C produced a high enough higher heating value (HHV) without excess energy input, which allowed for enough heat integration and lowered hydrochar drying costs as samples were more hydrophobic and had lower mass yields. Most electrical costs came from pelletization and reactor feed pumping, while most thermal costs came from HTC reaction heating. HTC reaction heating input was significantly larger for lower dry biomass-to-water ratio as there is less energy efficiency from heating more water content, which is why the higher ratio of 0.19 yielded better optimization. Overall pellet production cost was \$171.1 per ton of hydrochar and overall breakeven cost was equal to \$218 per ton; these costs were considered competitive with the cost of wood pellets (\$163.5–218 per ton) but not with coal costs. Co-treatment can remove pelletization costs as coal is not pelletized for traditional coal firing and heat integration is essential for viability. Shorter retention times have shown to be effective for hydrochar conversion and can increase revenue by allowing more product to be produced. To the author's knowledge, process economics have not been performed for a Co-HTC system, regardless of the feedstocks treated. Therefore, the main objective of this study is to evaluate techno-economic viability of Co-HTC of bituminous coal with miscanthus, while taking into consideration of capital costs and overall operating and maintenance costs over the plant lifetime. Heat integration will also be taken into consideration to reduce energy costs and to improve overall plant design.

2. Materials and Methodology

2.1. Co-Hydrothermal Carbonization Experiments

Detailed descriptions of the HTC and Co-HTC materials and methodologies for batch experiments can be found in a previous publication by Saba et al., [23]. *Miscanthus* (Miscanthus × giganteus) and Clarion Coal #4a (bituminous) were used for base case hydrothermal upgrading as well as Co-HTC. Base case HTC experiments consisted of loading a Parr stirred-batch reactor at 1:10 solids-to-DI water ratio, heating the contents to 230 °C, holding the reaction temperature isothermally for 30 min, and then cooling the contents down to room temperature with an ice-bath. Average heating time was 20 min to reach 5 °C below target temperature, and cooling times were less than 5 min to reach 80 °C. Same process was performed for Co-HTC runs, where a 1:1 miscanthus-to coal-mixing ratio was used. All products were dried in an oven at 105 °C for 24 h. Physio-chemical properties of hydrochars produced from HTC and Co-HTC experiments are shown in Table 1.

Table 1. Carbon, sulfur, and ash content of untreated feedstocks and hydrochars on a dry-basis.

Feedstock	Mass Yield (%)	HHV (MJ·kg^{-1})	Carbon (%)	Hydrogen (%)	Nitrogen (%)	Oxygen * (%)	Sulfur (%)	Ash (%)
Coal-untreated	-	27.4 ± 0.0	63.8 ± 1.2	4.1 ± 0.1	1.5 ± 0.1	26.0 ± 0.5	4.6 ± 0.8	11.1 ± 0.6
Miscanthus-untreated	-	19.7 ± 0.2	48.7 ± 0.1	5.7 ± 0.1	<0.5	45.4 ± 0.1	<0.5	1.3 ± 0.1
Coal 230 °C	99.2 ± 1.2	29 ± 1.0	69.2 ± 1.6	4.6 ± 0.3	1.7 ± 0.1	20.8 ± 1.9	3.8 ± 0.1	9.5 ± 1.1
Miscanthus 230 °C	57.2 ± 0.3	24.6 ± 0.4	59.3 ± 1.5	5.4 ± 0.1	<0.5	35.0 ± 1.5	<0.5	<0.5
Blend 230 °C	66.9 ± 0.8	26.1 ± 0.6	67.2 ± 2.0	4.2 ± 0.2	1.1 ± 0.1	25.6 ± 1.8	2.0 ± 0.3	5.3 ± 2.3

(* calculated by difference by source) [23].

2.2. Co-Hydrothermal Carbonization Process Overview

A process flow diagram (PFD) representing the Co-HTC system that treats miscanthus and coal is shown in Figure 1. This treatment includes filtering and drying the solid product that would be sold to coal fired power plants. In this process, miscanthus and coal are mixed, pressurized, and sent into a preheater/heat exchanger. Increasing the pressure prior to the pre-heater is necessary to avoid vaporization of the water. The slurry is pumped again to an outlet pressure of 34.5 bar (500 PSIG) for reactor operation and major and minor losses that are bound to occur in the system. Next, the slurry is sent into the reactor, which operates at 230 °C and 28.2 bar (410 PSIG) and is treated for a residence time of 5 min. All the heat exchangers and pumps are assumed to operate at an 80% efficiency in this model. After the residence time, the upgraded Co-HTC solid product, carbon dioxide gas, and process liquids are sent through the preheater to recover heat. Excess pressure is utilized to drive the product stream through a leaf filter; the hydrochar then is assumed to have a 20% moisture content after filtration [29]. From here, the solid product is sent to dry, where moisture content is brought down to 11%. This is well within the typical range of moisture content used for coal feeds at pulverized coal power plants [27]. The solid Co-HTC hydrochar is now ready to be utilized at the power plant. The gas is vented into the atmosphere and process liquid is sent to a wastewater treatment plant. Miscanthus feedstock and dried hydrochar will be stored onsite in separate storage tanks.

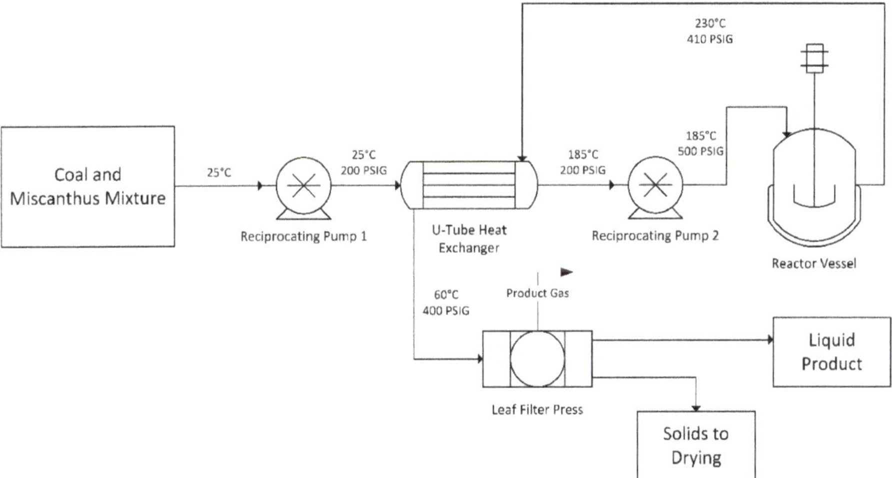

Figure 1. Simplified process flow diagram of Co-Hydrothermal Carbonization (Co-HTC) treatment used for economic analysis.

2.3. Co-HTC Process Economics

The methodology described below in this subsection is used for economic analysis. An economic analysis was performed for a scaled-up Co-HTC process to provide solid fuel for power plant combustion, such as an existing pulverized coal power plant. Capital cost estimations followed study estimates and preliminary design estimates, summarized by Turton et al., which were approximate parallels to estimate Class 2 and Class 3 in the Association for the Advancement of Cost Engineering Cost Classification System recommended practices report [30,31]. Following Turton et al., methodology, capital is estimated to fall between a −25%/+40% accuracy range. All prices shown have been adjusted to a 2016 USD standard (CEPCI = 541.7) unless otherwise noted; all prices in $/ton are in dry ton of fuel.

Data obtained from previous batch studies and analyses performed for Co-HTC hydrochars, produced at 230 °C, were applied towards mass and energy balances in order to determine energy outputs, unit operation electrical requirements, and product produced for combustion. Energy demands/outputs, electrical demands, and flow rates were then be used to determine investments costs (capital, sizing, etc.) and cash flows.

Investment cost and cash flows are used to determine net present value (NPV) of the plant via Equation (1) [30].

$$NPV = TCI + \sum_{k-1}^{n} F_k \cdot (1+i)^{-k} \tag{1}$$

where TCI is total capital investment, F_k is the annual after-tax cash flow, i is the interest rate, k is the year being evaluated using a year zero baseline, and n is the total number of years the plant is operating. NPV is used to find the product selling price for the plant to breakeven; this is done by setting NPV equal to zero and solving for the price of product. It should also be noted that a positive NPV indicates plant profit and project feasibility while a negative NPV indicates plant losses.

Capital costs for the plant were determined using the module costing technique [30]. Individual equipment base costs were calculated via Equation (2) [30].

$$\log(C_p^\circ) = K_1 + K_2 \cdot \log(A) + K_3 \cdot (\log(A))^2 \tag{2}$$

where, $C^\circ{}_p$ is the base cost of equipment made from carbon steel and designed for ambient pressure. K_1, K_2, and K_3 are constants found from Turton et al., [30] and are unique for each unit operation. A is the primary design parameter that is determined via mass balances, energy balances, and respective equipment design equations. Table 2 the K values for the units in the PFD.

Table 2. Constants used for equipment costs used in process flow diagram.

Model Unit	Qty	Units of A	K_1	K_2	K_3	Cost Modifier (F_{BM})
Jacketed Agitated Reaction Chamber	1	m^3	4.1052	0.532	−0.0005	4.00
Reciprocating pump (Pre)	1	kW	3.8696	0.3161	0.122	5.66
Reciprocating pump (Post)	1	kW	3.8696	0.3161	0.122	6.51
U-tube heat exchanger	1	m^2	4.1884	−0.2503	0.1974	4.59
Leaf filter press	1	m^2	3.8187	0.6235	0.0176	1.80
API fixed roof tank (miscanthus storage)	1	m^3	4.8509	−0.3973	0.1445	1.00
API fixed roof tank (product storage)	1	m^3	4.8509	−0.3973	0.1445	1.00

Once $C^\circ{}_p$ is calculated, it is multiplied by the cost modifier (F_{BM}) to determine the bare module cost (C_{BM}) of the equipment, as shown in Equation (3). F_{BM} incorporates not only material and pressure rating modifiers, but also indirect and direct costs for C_{BM}. Direct costs include equipment cost from the manufacturer and includes various costs ranging from piping and electric costs to labor costs. Indirect costs account from items such as freight costs, construction overhead, and additional engineering expenses. The overall fixed capital investment (FCI) is then calculated by summing up individual bare module costs. FCI also includes general fees, unforeseen costs (i.e., contingency costs), and supporting site costs (i.e., auxiliary costs). From previous literature, these additional costs are approximated as 3%, 15%, and 50% of C_{BM}, respectively; FCI is represented by Equation (4) [32].

$$C_{BM} = C_p^\circ \cdot F_{BM} \tag{3}$$

$$FCI = \sum_{j}^{m} C_{BM\,j} + 0.03\sum_{j}^{m} C_{BM\,j} + 0.15\sum_{j}^{m} C_{BM\,j} + 0.5\sum_{j}^{m} C_{BM\,j} \tag{4}$$

As shown in Equation (5), FCI is added with working capital (WC) in order to determine TCI. WC is the capital cost associated with the early phases of plant start-up and was approximated by Equation (6) [32]. It is important to note that the WC investment is not depreciated and is also recovered in the plant's cash flow in the final year "n".

$$\text{TCI} = \text{FCI} + \text{WC} \tag{5}$$

$$\text{WC} = 0.1 \cdot (\text{FCI} + C_{OL} + C_{RM}) \tag{6}$$

where, C_{OL} and C_{RM} are operating labor costs and raw material costs, respectively. Where "raw materials" are the miscanthus and bituminous coal used for Co-HTC. F_k will be determined by summing up the after-tax net profit (ANP) and depreciation (d) for the year (k). ANP and F_k are determined with Equations (7)–(9) [30].

$$F_k = \text{ANP} + d \tag{7}$$

$$\text{ANP} = \text{NP} - t \cdot \text{NP} \tag{8}$$

where NP is the net profit and t is the tax rate. Multiplying NP by t gives the profit/income that is taxed, so subtracting that quantity from NP itself yields ANP. NP is further defined as,

$$\text{NP} = R - \text{COM}_d - d \tag{9}$$

Equation (9) shows that NP is determined by subtracting expenses from revenue (R) which is represented by the cost of manufacturing (COM_d) and d. Equations (7)–(9) are combined to form Equation (10) [30].

$$F_k = (R - \text{COM}_d - d) - t \cdot (R - \text{COM}_d - d) + d \tag{10}$$

Equation (10) can then be simplified into Equation (11).

$$F_k = (R - \text{COM}_d - d)(1 - t) + d \tag{11}$$

COM_d is calculated with Equation (12) from general plant upkeep costs, C_{OL}, utility costs (C_U), C_{WT}, and C_{RM}. General plant maintenance is assumed to be an 18% the cost of the initial FCI.

$$\text{COM}_d = 0.18\text{FCI} + 2.73C_{OL} + 1.23(C_U + C_{WT} + C_{RM}) \tag{12}$$

A multiplier is used for C_{OL} in order to also account for administrative, supervisory, or laboratory costs. The multipliers for utilities, waste treatment, and transportation additionally account for any fluctuating or indirect costs. C_{OL} is sometimes referred to as fixed operating and maintenance (O&M) costs; C_U, C_{WT}, and C_{RM} are sometimes referred to as variable O&M costs. O&M parameters were summarized in Table 3. R is determined by multiplying the product produced each year by an estimated product price. The breakeven product is solved for by an incrementally changing the product price until the object function "NPV = 0" is satisfied through Equation (1). N_{OL} is the base number of operators per shift and is determined using Equation (13) [30].

$$N_{OL} = \left(6.29 + 31.7 \cdot P^2 + 0.23 \cdot N_{np}\right)^{0.5} \tag{13}$$

where, P is the total number particulate handling unit operations and N_{np} is the total number of non-particulate handling unit operations. Salary was reported from INL economic analysis and the C_{OL} multiplier allows for calculation of additional costs associated outside operator labor.

Table 3. Fixed and variable operating and maintenance (O&M) parameters and economic parameters used for estimating operating labor and costs associated with manufacturing.

Fixed O&M Parameters			**Variable O&M Parameters**		
C_{OL} multiplier	2.76	[30]	Variable O&M multiplier	1.23	[30]
Operator salary ($/year)	52,700	[33]	Miscanthus ($/ton)	38	[34]
N_{np}	2	-	Bituminous coal ($/ton)	53.24	[35]
P	1	-	Water utility ($ m^{-3})	1.12	[36]
Op Labor	28	-	Waste water disposal ($/ton)	0.74	[37]
			Electricity cost ($ kWh^{-1})	0.066	[38]
			Cost of natural gas ($ m^{-3})	0.132	[39]
			Natural gas energy content (MJ m^{-3})	38.64	[40]
Economic Parameters					
Tax rate (%)	25	[41]	CRF	0.12	-
Annual interest rate (%)	10	[27,42]	Salvage value	0	[32]
Plant life (years)	20	-	Depreciation	7-year MACRS	[27,42]
Streaming factor (SF)	0.90	[33]	2016 CEPCI	541.7	[43]

2.4. Cost of Electricity

General economic parameters for determining NPV, capital investments, and cash flows are presented in Table 4. Most parameters have been assumed through NREL and NETL case study reports [27,42,44]. CRF is the capital recovery factory (Equation (14) [30]) which is used to calculate the total cost of electricity (COE). COE provides another parameter for evaluating plant costs and is calculated using Equations (15)–(18). CRF is used in Equation (15) to convert capital investments over the plant's lifetime into an annuity.

$$\mathrm{CRF} = \frac{\mathrm{i}(1+\mathrm{i})^{\mathrm{n}}}{((1+\mathrm{i})^{\mathrm{n}}-1)} \tag{14}$$

$$\mathrm{COE_C} = \frac{\mathrm{CRF}\cdot(\mathrm{FCI}+\mathrm{WC})}{\mathrm{SF}\cdot\mathrm{Plant\_Capacity}} \tag{15}$$

$$\mathrm{COE_F} = \frac{0.18\cdot\mathrm{FCI}+2.76\cdot\mathrm{C_{OL}}}{\mathrm{SF}\cdot\mathrm{Plant\_Capacity}} \tag{16}$$

$$\mathrm{COE_V} = \frac{1.23\cdot(\mathrm{C_U}+\mathrm{C_{WT}})}{\mathrm{SF}\cdot\mathrm{Plant\_Capacity}} \tag{17}$$

$$\mathrm{COE_{RM}} = \frac{1.23\cdot\mathrm{C_{RM}}}{\mathrm{SF}\cdot\mathrm{Plant\_Capacity}} \tag{18}$$

where $\mathrm{COE_C}$ is the capital COE, $\mathrm{COE_F}$ is the fixed operating COE, $\mathrm{COE_V}$ is the variable COE, and $\mathrm{COE_{RM}}$ is the raw material COE (sometimes referred to as fuel costs in literature). These separate costs are summed to form Equation (19), modified from literature [45].

$$\mathrm{COE} = \mathrm{COE_C}+\mathrm{COE_F}+\mathrm{COE_V}+\mathrm{COE_{RM}} \tag{19}$$

Table 4. Experimental data and flowrates used for overall material and energy balances for equipment design.

Parameters	Coal	Miscanthus	Blend
Plant rating (MWe)		110	
Thermal to electric efficiency (%)		35	
Reaction temperature (°C)		230	
Coal: Miscanthus solids ratio	1:0	0:1	1:1
Total feed water content (%)		70	
Gas production ($\times 10^{-3}$ kg/kg solid Feed)	1.25	3.7	4.9
Total process feed (kg hr^{-1})	131,363	268,025	215,993
Total solid feed (kg hr^{-1})	39,409	80,407	64,798
Total water feed (kg hr^{-1})	91,954	187,617	151,195
Produced hydrochar on dry-basis (kg hr^{-1})	39,015	45,993	43,350
Total process liquid after treatment (kg hr^{-1})	92,999	221,734	172,326
Process liquid sent to wastewater treatment (kg hr^{-1})	82,545	210,236	161,488
Moisture dried from post-filter hydrochar (kg hr^{-1})	4932	5814	5480

2.5. Sensitivity Analysis

Sensitivity analysis was performed for the base case economic model to evaluate changes in the breakeven selling price. An individual process/economic parameter was changed, while all other independent process/economic parameters were held constant in order to evaluate the change in breakeven selling price. The individual parameter was changed to a lower sensitivity bound (LSB) and a higher sensitivity bound (HSB) with respect to the base case scenario; resulting breakeven prices were then plotted on a sensitivity chart (also known as a tornado plot). It should be noted that the determination of the LSB and HSB are based on previously observed market fluctuations (e.g., utility costs) or when not available, a general overall change in the parameter was assumed (e.g., FCI).

3. Results and Discussion

3.1. Plant Processing for Coal, Miscanthus, and Blend Hydrothermal Treatment

Hydrothermal processing feed rates were determined for coal, miscanthus, and 50:50 blend by using a 110 MWe power plant basis, assuming a thermal to electric efficiency of 35%, and applying experimentally determined batch reactor mass and energy yields at 230 °C (Table 4). The water content for treatment was reduced to 70% as solids to liquids loading is tertiary to process parameters temperature and residence time; complete feedstock submergence by the liquid phase at reaction temperature (i.e., volume of the liquid water is greater than the volume of the solid) is necessary for it to be considered undergoing HTC [46,47]. A miscanthus-only treatment facility would process more than twice the total feed of coal-only treatment as a result of the low miscanthus mass yield (57.2%) and the high coal mass yield (nearly 99%). Since the low miscanthus yield requires higher untreated biomass loading, the amount of water needed for treatment also increases; nearly three times the process liquid is sent for wastewater treatment at 210,236 kg hr^{-1} for miscanthus-only treatment than the coal-only treatment. The total solid feed into the reactor for Co-HTC that would meet the power requirement was determined to be 64,798 kg hr^{-1}, producing a hydrochar flowrate of 43,350 kg hr^{-1}. The synergistic interaction during Co-HTC causes the higher production of gas flow at 318 kg hr^{-1} despite coal-only and miscanthus-only gas flowrates of 49 and 298 kg hr^{-1}, respectively [23]. Co-HTC process flow conditions and mass balances for the solid, liquid, and gas inputs/outputs were used to approximate design parameters and equipment sizing, as economics analysis of Co-HTC was the objective of this study. Only one of each piece of equipment shown in Figure 1 and summarized in Table 2 was used for material processing and meeting power demand. Sizing, loading, and equipment quantity were then used to estimate capital costs.

3.2. Estimation of Capital Costs

Table 5 provides the capital cost analysis for the plant using 2016 pricing and C_{BM} and TCI distribution are presented. The total estimated C_{BM} for the plant is $5.30 million where pumping accounts for 45.9% of the total costs, followed by the heat exchanger and leaf filter contributing to ~21% of the total cost, each. Lastly, material storage and the reactor system accounted for 5 to 7.5% of C_{BM}. Material of construction used to calculate part of F_{BM} was SS-316, thus F_{BM} from the material was standard for comparison among all the equipment other than storage. SS-316 was used for the material of construction since process conditions are not corrosive enough to require more resistant and more expensive alloys, such as nickel-based alloys, for construction. Pressure factor contributions to F_{BM}, however, varied more as the heat exchanger and post-heat exchanger pump operate at higher operating pressures and temperatures. The leaf filter shared an unexpected higher amount of the C_{BM} cost compared to the other unit operations. However, the filter is limited by filter area and using a film thickness of 0.03 m [48] and a clearing rate of 0.25 hr^{-1}, can still only remove so much of the hydrochar; the lower filtration time is contributed to the hydrophobic nature of the hydrochar. Meanwhile, FCI, scaled from C_{BM} and WC then contribute to a TCI of $12.27 million. Breakeven cost will not be significantly affected by TCI since the quantity for each equipment is one. Though only one unit is used, contingent units are priced for in Equation 4. High manufacturing costs will contribute to a higher selling price, discussed in the next section as well as in the sensitivity analysis.

Table 5. Cost summary of Co-HTC plant producing fuel for 110 MWe coal power plant.

Capital Costs			Manufacturing Cost		
Item	**Cost**	**Cost Distribution**		**$/year**	**$/ton**
Reactor system	$396,246	7.50%	Estimated FCI upkeep	1,601,782	4.7
Pumping systems	$2,431,594	45.90%	Labor costs	4,085,936	12.0
Heat recovery system	$1,100,931	20.80%	Utilities	3,413,468	10.0
Solid product filtration and dewatering	$1,092,394	20.60%	Waste treatment	1,158,841	3.4
Storage	$275,762	5.20%	Raw materials	28,666,025	83.9
C_{BM}	$5,297,000	-	**Total**	**38,926,052**	**113.9**
FCI	$8,898,792	-			
WC	$3,370,118	-			
TCI	$12,268,910	-			
			Operating and Maintenance Cost		
COE ($ MWh^{-1})			*Fixed O&M costs*		
Capital COE	1.66		C_{OL} ($/year)	1,496,680	
Fixed operating COE	6.61		*Variable O&M costs*		
Variable operating COE	5.27		C_U ($/year)	2,775,177	
Raw material (fuel cost) COE	33.05		C_{WT} ($/year)	942,147	
Total COE	**46.60**		C_{RM} ($/year)	23,305,712	

3.3. Estimation of Manufacturing Costs

O&M base costs were determined via costs and parameters provided in Table 5. These base costs are then multiplied by their respective multipliers to account for indirect costs or miscellaneous costs (discussed earlier) and are incorporated with FCI upkeep in order to determine yearly manufacturing costs (shown in Table 5). Total manufacturing costs are $38.93 million per year, the bulk of those costs arise from purchasing raw materials for upgrading which comes out to $28.67 million per year. FCI upkeep and waste treatment account for 3–4% of the total manufacturing costs while labor costs and utilities makeup nearly 9–10% of the cost. The remaining 73% make up the cost of miscanthus and coal which convert to a cost of $83.9 to produce one ton of hydrochar product. Wirth et al., similarly saw that biomass supply costs for an HTC plant contributed to 37–59% of costs and was more significant than the cost of biomass-to-hydrochar conversion [49]. The manufacturing costs associated with raw materials are expected as purchasing costs for miscanthus and coal are $38 and $53.24 per ton. Total manufacturing cost is also equivalent to $113.9 per ton of hydrochar, giving initial insight into the breakeven cost, before solving Equation (1). The selling price of the produced hydrochar must be greater than $113.9 per ton in order to break even.

3.4. Estimation of Cost of Electricity (COE)

The total COE (Table 5) represents the plant's first year total cost of energy generation per total energy supplied annually. As observed with total manufacturing cost itemization, COE_{RM} accounts for most of the total COE. The COE for Co-HTC plant should be compared with the fuel COE from different power plants presented in NETL techno-economic analyses, rather than the total COE. This is important as Co-HTC is a means of upgrading waste or low value feedstocks into a combustible fuel, which will then be burned at a power plant. Thus, the total COE of the Co-HTC plant should only be compared to COE costs associated with fuel for the power plant in existing literature. The Co-HTC COE comes out to $46.60 MWh^{-1} for a plant operating at 110 MWe, whereas the fuel costs for pulverized coal using subcritical or supercritical boiler technology range from $22.8 to $29.8 MWh^{-1} (converted from to $2011 $2016) for plant capacities of 550 MWe [27]. These Co-HTC prices range from 56.4–105 % more than the base pulverized coal costs at approximately 20% of the net, plant capacity. The original COE for these coal plants range from $82.0 to $133.5 MWh^{-1}, where fuel costs are second to capital costs in terms of total makeup.

3.5. Breakeven Price and Sensitivity Analysis

With TCI and total manufacturing costs calculated, breakeven selling price were determined by setting NPV in Equation (1) equal to zero. Breakeven price for the baseline scenario using parameters in Tables 3 and 4 was determined to be $117.91 per ton of hydrochar. At $4.49 GJ^{-1}, the price of Co-HTC hydrochar is 2.30 times more than the average price of bituminous coal ($1.95 GJ^{-1}) and 1.31 times more expensive that the cost of natural gas ($3.42 GJ^{-1}) on an energy content basis. The high selling price for the hydrochar is in accordance with the high manufacturing costs and COE discussed earlier as miscanthus and coal purchase cost were $38 and $53.24 per ton of respective feedstock. Since selling price was impacted significantly by current feedstock prices, the change in selling price was evaluated by assuming different raw material purchasing costs in a sensitivity analysis. Additionally, other process parameters were increased and decreased to observe their contribution to the selling price. Table 6 provides all parameters altered for the sensitivity analysis; the baseline parameters are presented with the ranging sensitivity values as well as the resulting selling prices. The percent change in breakeven costs with respect to the baseline breakeven cost can be found in Figure 2. Hydrochar throughput and raw material prices have a more significant effect on final selling prices than manufacturing related costs. The latter was previously observed in previous sections, as breakeven prices can decrease by 14% by lowering feedstock price or can increase by approximately 10%. This variation indicates that co-treatment costs can improve by utilizing cheaper biomass options such as waste feedstocks, which have been previously shown to be upgraded via HTC [17]. Similarly, lower rank coal (e.g., lignite) fuel properties can benefit from HTC treatment [50,51] and can be purchased at lower costs [35].

Table 6. Parameter ranges for sensitivity analysis used to determine breakeven cost ranges.

Items	Baseline Scenario	Sensitivity Range	LSB Breakeven Cost ($/ton)	HSB Breakeven Cost ($/ton)
Baseline scenario breakeven price ($/ton)	117.24	-	-	-
FCI ($ 10^6)	8.90	8.01–10.23	116.44	118.44
Hydrochar yield (%)	66.9	60–74	130.58	106.1
Hydrochar HHV (MJ·kg^{-1})	26.1	24–29.5	116.03	119.09
Miscanthus price ($/ton)	38	20–50	100.69	128.27
Coal price ($/ton)	53.24	45.4–66.7	110.00	129.58
Waste treatment ($/ton)	0.74	0.5–2.0	116.14	123.01
Water utilities ($ m^{-3})	1.12	0.66–2.11	115.27	121.52
Cost of natural gas ($ m^{-3})	0.132	0.092–0.300	115.85	123.1
Power rating (MWe)	110	55–550	130.85	106.88

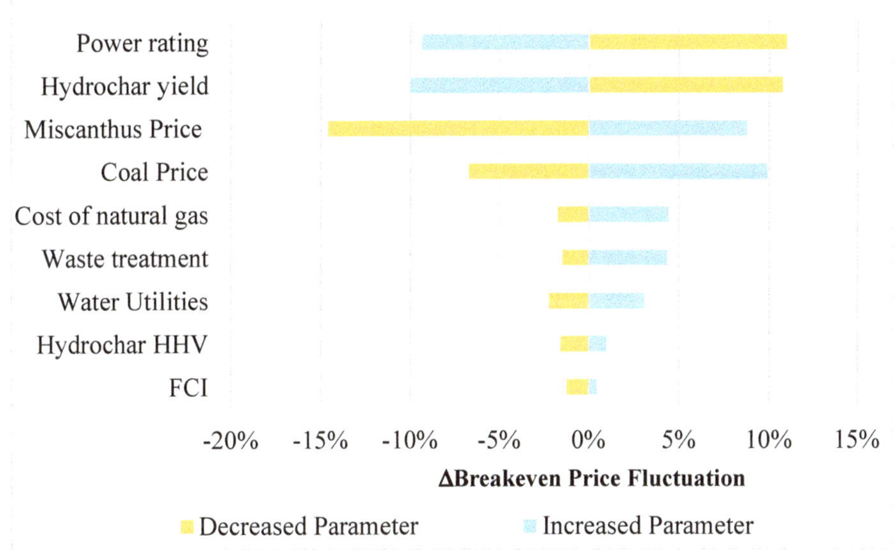

Figure 2. Sensitivity analysis showing change in breakeven price of product with respect to baseline breakeven cost.

Retaining more of the reactor feed as hydrochar lowers breakeven price as there is more product to sell. If reaction yield is increased by just 4%, breakeven price drops down to $106.10 per ton from the base case cost of $117.24 per ton. In the same form of price impact, increasing plant production (i.e., increasing feed flowrate) by increasing power rating demand produces more hydrochar. As expected, this lowers breakeven price as there is more product to sell. Scaling down power demand to 55 MWe, requires less hydrochar production for burning and consequently increases the base cost to $130.59 per ton while a power rating of 550MWe plant decreases the selling price by 9% to $106.88 per ton. Changing power demand for sensitivity analysis changes other costing parameters (quantity of equipment, equipment sizing parameters, utility usage, labor, etc.) since reactor throughput increases significantly. Thus, breakeven price was evaluated for different power demand loads (Figure 3). Breakeven price decreases with loading capacity, as is expected with economy of scale, and begins to plateau at 220 MWe.

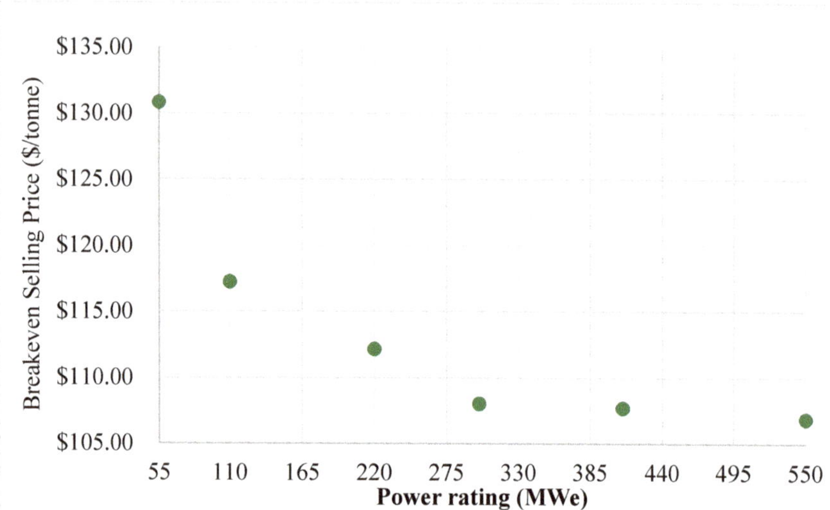

Figure 3. Breakeven selling price for hydrochar at different power ratings.

Although a power rating of 550 MWe and 55 MWe have similar breakeven costs to hydrochar yields of 74% and 60%, the impact from hydrochar yield is much more impactful in changing breakeven costs. The plant production for a power rating of 550 MWe produces 1.71×10^6 tons of hydrochar per year, while plant production for the 110 MWe load at 74% mass yield produces only 3.78×10^5 tons of hydrochar per year. Both cases are approximately $107 per ton despite an order of magnitude difference in production. Even at 300 MWe, where breakeven price in Figure 3 starts to steady, 9.32×10^5 tons of hydrochar per year are produced. Economy of scale allows for reduced breakeven cost as the plant is scaled up (observed with power demand), however the increase in equipment capital and increase in manufacturing cost offsets additional revenue created from increased hydrochar production. Overall, biomass to coal ratio can be adjusted to produce higher yields and feedstock flowrate can be increased for more hydrochar processing, however, the latter should not be done where additional equipment and manufacturing costs increase to the point of counterbalancing the growth in profit.

4. Conclusions

A Co-HTC process producing 43,350 kg hr^{-1} was designed to deliver fuels for a 110 MWe power plant using Clarion coal #4a and miscanthus with 50:50 (wt % dry basis) blend. Total capital investment was estimated at $12.7 million, where pumping was the predominant capital investment followed by heat exchangers and filtration. Total manufacturing cost is $38.9 million per year and equivalent to $113.9 per ton of product produced, indicating that the product selling price cannot be lower than this cost to breakeven. The cost of electricity for making Co-HTC hydrochar is around $46 per MWh^{-1}, nearly twice the cost of the fuel costs associated with standard coal-fired plants. The breakeven selling price of the produced hydrochar was $117.24 per ton, where the cost of purchasing feedstocks for upgrading accounted for most of the breakeven price. Sensitivity analysis showed that power rating, cost of coal and miscanthus, and hydrochar yield could reduce this breakeven selling price significantly. Attempting to treat co-treat abundant waste feedstocks and lower ranks coals via HTC can provide lower breakeven costs as these raw materials can be purchased at cheaper prices and should be considered for future studies.

Author Contributions: A.S. performed the experiments, analyzed the results and wrote the article draft, K.M. examined the techno-economic model, and M.T.R. reviewed the article.

Funding: This research is funded by Ohio Coal Development Office project no R-17-05.

Acknowledgments: The authors acknowledge Ravi Garlapalli and Sarah Davis from Ohio University for providing coal and miscanthus samples, respectively, and for providing initial characterization of the feedstocks. The authors are also thankful to Mr. Nepu Saha at the Institute for Sustainable Energy and the Environment (ISEE) for their meaningful discussions in this project.

Conflicts of Interest: The authors declare no conflicts of interest.

List of Initialisms/Acronyms

SF	Streaming factor
MACRS	Modified accelerated cost recovery system
ANP	After-tax net profit
C_{BM}	Bare module cost
COE	Cost of electricity
Co-HTC	Co-Hydrothermal Carbonization
C_{OL}	Operating labor costs
COM_d	Cost of manufacturing without depreciation
CRF	Capital recovery factor
C_{RM}	Raw material (or fuel) costs
C_U	Utility cost
C_{WT}	Waste treatment cost
d	Depreciation
EOIR	Energy output to input ratio
F_{BM}	Bare module cost modifier
FCI	Fixed capital investment
F_k	Annual after-tax cash flow
HHV	Higher heating value
HSB	Higher sensitivity bound
HTC	Hydrothermal Carbonization
i	Annual interest rate
k	Evaluation year
LSB	Lower sensitivity bound
N_{NP}	Total number of non-particulate handling unit operations
N_{OL}	Base number of operators per shift
NP	Net profit
NPV	Net present value
O&M	Operating and maintenance
P	Total number particulate handling unit operations
t	Tax rate
TCI	Total capital investment
WC	Working capital

References

1. United States Environmental Protection Agency (USEPA). *Inventory of U.S. Greenhouse Gas. Emissions and Sinks: 19902–016*; USEPA: Washington, DC, USA, 2018; pp. 2–37.
2. Skone, T.; Littlefield, J.; Eckard, R.; Cooney, G.; Wallace, R.; Marriott, J. *Role of Alternative Energy Sources: Pulverized Coal and biomass Co-Firing Technology Assessment*; DOE/NETL: Pittsbourgh, PA, USA, 2012.
3. Administration, U.S.E.I. *Annual Energy Outlook 2016: With Projections to 2040*; US Energy Information Administration, Office of Energy Analysis, US Department of Energy: Washington, DC, USA, 2016.
4. U.S. Department of Energy. *Billion Ton Update: Biomass Supply for a Bioenergy and Bioproducts Industry*; Oak Ridge National Laboratory: Oak Ridge, TN, USA, 2016.
5. Energy Information Administration (EIA). *Monthly Energy Review November 2018*; Office of Energey Statistics: Washington, DC, USA, 2018.
6. Mun, T.Y.; Tumsa, T.Z.; Lee, U.; Yang, W. Performance evaluation of co-firing various kinds of biomass with low rank coals in a 500 MWe coal-fired power plant. *Energy* **2016**, *115*, 954–962. [CrossRef]

7. Manzanera, M. *Alternative Fuel*; IntechOpen: London, UK, 2011; ISBN 978-953-307-372-9.
8. Smoot, L.D.; Smith, P.J. *Coal Combustion and Gasification*; Springer: New York, NY, USA, 2013.
9. Vamvuka, D.; Kakaras, E. Ash properties and environmental impact of various biomass and coal fuels and their blends. *Fuel Process. Technol.* **2011**, *92*, 570–581. [CrossRef]
10. Bownocker, J.A.; Department of the Interior. *The Coal Fields of the United States*; Government Printing Office: Washington, DC, USA, 1971.
11. Langholtz, M.H.; Stokes, B.J.; Eaton, L.M. *2016 Billion-Ton Report: Advancing Domestic Resources for a Thriving Bioeconomy*; DOE/EE-1440; Other: 7439 United States 10.2172/1271651; EERE Publication and Product Library: Washington, DC, USA, 2016.
12. McKendry, P. Energy production from biomass (part 1): Overview of biomass. *Bioresour. Technol.* **2002**, *83*, 37–46. [CrossRef]
13. Zulfiqar, M.; Moghtaderi, B.; Wall, T.F. Flow properties of biomass and coal blends. *Fuel Process. Technol.* **2006**, *87*, 281–288. [CrossRef]
14. *Biomass Cofiring in Coal-Fired Boilers*; U.S. Dept. of Energy, Office of Energy Efficiency and Renewable Energy: Washington, DC, USA, 2004. Available online: http://purl.access.gpo.gov/GPO/LPS99144 (accessed on 3 January 2019).
15. Favas, G.; Jackson, W.R. Hydrothermal dewatering of lower rank coals. 1. Effects of process conditions on the properties of dried product. *Fuel* **2003**, *82*, 53–57. [CrossRef]
16. Kruse, A.; Zevaco, A.T. Properties of hydrochar as function of feedstock, reaction conditions and post-treatment. *Energies* **2018**, *11*, 674. [CrossRef]
17. Román, S.; Libra, J.; Berge, N.; Sabio, E.; Ro, K.; Li, L.; Ledesma, B.; Álvarez, A.; Bae, S. Hydrothermal carbonization: Modeling, final properties design and applications: A review. *Energies* **2018**, *11*, 216. [CrossRef]
18. Yu, Y.; Liu, J.; Wang, R.; Zhou, J.; Cen, K. Effect of hydrothermal dewatering on the slurryability of brown coals. *Energy Convers. Manag.* **2012**, *57*, 8–12. [CrossRef]
19. Zhang, B.; Heidari, M.; Regmi, B.; Salaudeen, S.; Arku, P.; Thimmannagari, M.; Dutta, A. Hydrothermal carbonization of fruit wastes: A promising technique for generating hydrochar. *Energies* **2018**, *11*, 2022. [CrossRef]
20. Shen, Y.; Yu, S.; Ge, S.; Chen, X.; Ge, X.; Chen, M. Hydrothermal carbonization of medical wastes and lignocellulosic biomass for solid fuel production from lab-scale to pilot-scale. *Energy* **2017**, *118*, 312–323. [CrossRef]
21. Zhang, X.; Zhang, L.; Li, A. Hydrothermal co-carbonization of sewage sludge and pinewood sawdust for nutrient-rich hydrochar production: Synergistic effects and products characterization. *J. Environ. Manag.* **2017**, *201*, 52–62. [CrossRef] [PubMed]
22. Nonaka, M.; Hirajima, T.; Sasaki, K. Upgrading of low rank coal and woody biomass mixture by hydrothermal treatment. *Fuel* **2011**, *90*, 2578–2584. [CrossRef]
23. Saba, A.; Saha, P.; Reza, M.T. Co-Hydrothermal Carbonization of coal-biomass blend: Influence of temperature on solid fuel properties. *Fuel Process. Technol.* **2017**, *167* (Suppl. C), 711–720. [CrossRef]
24. Funke, A.; Ziegler, F. Heat of reaction measurements for hydrothermal carbonization of biomass. *Bioresour. Technol.* **2011**, *102*, 7595–7598. [CrossRef] [PubMed]
25. McGaughy, K.; Reza, M.T. Hydrothermal carbonization of food waste: simplified process simulation model based on experimental results. *Biomass Convers. Biorefinery* **2018**, *8*, 283–292. [CrossRef]
26. Unrean, P.; Lai Fui, B.C.; Rianawati, E.; Acda, M. Comparative techno-economic assessment and environmental impacts of rice husk-to-fuel conversion technologies. *Energy* **2018**, *151*, 581–593. [CrossRef]
27. Fout, T.; Zoelle, A.; Keairns, D.; Turner, M.; Woods, M.; Kuehn, N.; Shah, V.; Chou, V.; Pinkerton, L. *Cost and Performance Baseline for Fossil Energy Plants Volume 1a: Bituminous Coal (PC) and Natural Gas to Electricity*; Revision 3; National Energy Technology Laboratory Report; DOE/NETL-2015/1723; NETL: Washington, DC, USA, 2015.
28. Lucian, M.; Fiori, L. Hydrothermal carbonization of waste biomass: Process design, modeling, energy efficiency and cost analysis. *Energies* **2017**, *10*, 211. [CrossRef]
29. Kambo, H.S.; Dutta, A. Strength, storage, and combustion characteristics of densified lignocellulosic biomass produced via torrefaction and hydrothermal carbonization. *Appl. Energy* **2014**, *135*, 182–191. [CrossRef]
30. Turton, R.; Bailie, R.C.; Whiting, W.B.; Shaeiwitz, J.A. *Analysis, Synthesis and Design of Chemical Processes*; Pearson Education: London, UK, 2008.

31. AACE. *Cost Estimate Classification System As Applied in Engineering, Procurement, and Construction for the Process. Industries*; AACE International: Morgantown, WV, USA, 2016.
32. Mahmood, R.; Parshetti, G.K.; Balasubramanian, R. Energy, exergy and techno-economic analyses of hydrothermal oxidation of food waste to produce hydro-char and bio-oil. *Energy* **2016**, *102*, 187–198. [CrossRef]
33. Hu, H.Q.; Westover, T.L.; Cherry, R.; Aston, J.E.; Lacey, J.A.; Thompson, D.N. Process simulation and cost analysis for removing inorganics from wood chips using combined mechanical and chemical preprocessing. *Bioenerg. Res.* **2017**, *10*, 237–247. [CrossRef]
34. Jain, A.K.; Khanna, M.; Erickson, M.; Huang, H.X. An integrated biogeochemical and economic analysis of bioenergy crops in the Midwestern United States. *GCB Bioenerg.* **2010**, *2*, 217–234. [CrossRef]
35. Energy Information Administration (EIA). *Annual Coal Report 2017*; Energy U.S.D. o., Ed.; United States Energy Information Administration: Washington, DC, USA, 2018.
36. EERE. *Water and Wastewater Annual Price Escalation Rates for Selected Cities across the United States*; Energy U.S.D. o., Ed.; Office of Energy Efficiency & Rewnewable Energy: Washington, DC, USA, 2017.
37. Prieto, D.; Swinnen, N.; Blanco, L.; Hermosilla, D.; Cauwenberg, P.; Blanco, A.; Negro, C. Drivers and economic aspects for the implementation of advanced wastewater treatment and water reuse in a PVC plant. *Water Resour. Ind.* **2016**, *14*, 26–30. [CrossRef]
38. Energy Information Administration (EIA). *Electric Power Monthly with Data for August 2018*; United States Energy Information Administration: Independent Statistics & Analysis: Washington, DC, USA, Ocotober 2018.
39. Energy Information Administration (EIA). United States Natural Gas Industrial Price. Available online: https://www.eia.gov/dnav/ng/hist/n3035us3m.htm (accessed on 3 January 2019).
40. Energy Information Administration (EIA). What are Ccf, Mcf, Btu, and therms? How do I Convert Natural Gas Prices in Dollars per Ccf or Mcf to dollars per Btu or therm? Available online: https://www.eia.gov/tools/faqs/faq.php?id=45&t=8 (accessed on 3 January 2019).
41. Statnick, R. Ohio Coal Development Agency Site Visit and Economic Analysis Recommendations. In *Funding Agency Program Manager Tax Usage Reccomendation.*; State of Ohio Development Services Agency: Athens, OH, USA, 2018.
42. Davis, R.; Tao, L.; Tan, E.C.D.; Biddy, M.J.; Beckham, G.T.; Scarlata, C.; Jacobson, J.; Cafferty, K.; Ross, J.; Lukas, J.; et al. *Process Design and Economics for The Conversion Of Lignocellulosic Biomass to Hydrocarbons: Dilute-Acid and Enzymatic Deconstruction of Biomass to Sugars and Biological Conversion of Sugars to Hydrocarbons*; NREL: Denver, CO, USA, 2013.
43. Lozowski, D. The Chemical Engineering Plant Cost Index. Available online: https://www.chemengonline.com/pci-home (accessed on 3 January 2019).
44. Davis, R.D.; Kinchin, C.; Markham, J.; Tan, E.C.D.; Laurens, L.M.; Sexton, D.; Knorr, D.; Schoen, P.; Lukacs, J. *Process. Design and Economics for the Conversion of Algal Biomass to Biofuels: Algal Biomass Fractionation to Lipid and Carbohydrate-Derived Fuel Products*; National Renewable Energy Laboratory: Golden, CO, USA, 2014.
45. Fisher, J.C. *Introduction to Performing a Techno-Economic Analysis for Power Generation Systems*; NETL: Golden, CO, USA, 2015.
46. Makela, M.; Benavente, V.; Fullana, A. Hydrothermal carbonization of lignocellulosic biomass: Effect of process conditions on hydrochar properties. *Appl. Energ.* **2015**, *155*, 576–584. [CrossRef]
47. Funke, A.; Reebs, F.; Kruse, A. Experimental comparison of hydrothermal and vapothermal carbonization. *Fuel Process. Technol.* **2013**, *115*, 261–269. [CrossRef]
48. Kriegel, P. Plate and Frame Filter Press. *Ind. Eng. Chem.* **1938**, *30*, 1211–1213. [CrossRef]
49. Wirth, B.; Eberhardt, G.; Lotze-Campen, H.; Erlach, B.; Rolinski, S.; Rothe, P. Hydrothermal Carbonization: Influence of Plant Capacity, Feedstock Choice and Location on Product Cost. In Proceedings of the 19th European Biomass Conference & Exhibition, Berlin, Germany, 6–10 June 2011; pp. 2001–2010.
50. Ge, L.; Zhang, Y.; Xu, C.; Wang, Z.; Zhou, J.; Cen, K. Influence of the hydrothermal dewatering on the combustion characteristics of Chinese low-rank coals. *Appl. Therm. Eng.* **2015**, *90*, 174–181. [CrossRef]
51. Liao, J.J.; Fei, Y.; Marshall, M.; Chaffee, A.L.; Chang, L.P. Hydrothermal dewatering of a Chinese lignite and properties of the solid products. *Fuel* **2016**, *180*, 473–480. [CrossRef]

Article

Hydrothermal Carbonization Brewer's Spent Grains with the Focus on Improving the Degradation of the Feedstock

Pablo J. Arauzo *, Maciej P. Olszewski and Andrea Kruse

Department of Conversion Technologies of Biobased Resources, Institute of Agricultural Engineering, University of Hohenheim, Garbenstrasse 9, 70599 Stuttgart, Germany; maciej.olszewski@uni-hohenheim.de (M.P.O.); andrea_kruse@uni-hohenheim.de (A.K.)

* Correspondence: pabloj.arauzo@uni-hohenheim.de; Tel.: +49-711-459-24705

Received: 22 October 2018; Accepted: 15 November 2018; Published: 21 November 2018

Abstract: Hydrochar is a very interesting product from agricultural and food production residues. Unfortunately, severe conditions for complete conversion of lignocellulosic biomass is necessary, especially compared to the conversion of sugar compounds. The goal of this work is to improve the conversion of internal carbohydrates by application of a two-steps process, by acid addition and slightly higher water content. A set of experiments at different temperatures (180, 200, and 220 °C), reaction times (2 and 4 h), and moisture contents (80% and 90%) was performed to characterize the solid (high heating value (*HHV*), elemental) and liquid product phase. Afterwards, acid addition for a catalyzed hydrolysis reaction during hydrothermal carbonization (HTC) and a two-steps reaction (180 and 220 °C) were tested. As expected, a higher temperature leads to higher C content of the hydrochar and a higher fixed carbon (FC) content. The same effect was found with the addition of acids at lower temperatures. In the two-steps reaction, a primary hydrolysis step increases the conversion of internal carbohydrates. Higher water content has no significant effect, except for increasing the solubility of ash components.

Keywords: hydrothermal carbonization (HTC); brewer's spent grains (BSG); hydrochar; acid addition; two-steps carbonization

---

1. Introduction

The two major conversion pathways to produce energy from lignocellulosic biomass are based on biological and thermochemical processes [1]. Among the several advantages of thermochemical processes versus biological ones are that the former have shorter conversion times and higher robustness, while biological processes require an accurate control using microorganisms. Due to its versatility and flexibility, thermochemical conversion technologies receive special attention by process developers. Thermochemical conversions can be divided depending on the feedstock's characteristics in dry processes where the feedstock's moisture content is between 5 wt.% and 10 wt.%, such as torrefaction, pyrolysis, and gasification; and in wet processes, which focus on feedstock with moisture content between 60 wt.% and 90 wt.% [2]. These wet processes are hydrothermal carbonization (HTC), hydrothermal liquefaction (HTL), and hydrothermal gasification (HTG).

This study focuses on the HTC process. HTC is a promising technology for processing wastes and residues with high moisture content, aiming to produce carbonaceous material [3], called hydrochar to distinguish it from other carbonaceous materials, such as biochar or coal [4–6]. The operating temperature range of HTC is from 180 °C to 250 °C, in order to maximize solid production (66 wt.% dry basis) [4,7,8] and minimize gas and liquid/solved organic compounds yield [9]. At temperatures from 250 °C to 373 °C the process is called HTL, so liquid phase is the main product [10]. If gas phase is the desired product phase, usually the temperature has to be increased near to or above the critical point of water; in this case the process is called HTG or supercritical water gasification (SCWG) [11].

The chemical mechanisms involved during the HTC process involve a sequence of different reactions. These are essentially hydrolysis, dehydration, decarboxylation, aromatization and re-condensation reactions [12–15]. As most of them also occur simultaneously, the study of the process is quite complex. According to Kruse et al. [16], the reactions are divided on two major pathways. Firstly, the solid-to-solid conversion of original biomass takes place (here, the structure of the hydrochar has the morphology and structural elements of the initial feedstock). Secondly, the solvation of the intermediates from converted biomass in the aqueous phase, which is followed by polymerization [7,17]. This is a simplification, because in the case of biomass both processes occur. The dominant pathway depends on the structure and composition of biomass (Karayıldırım et al. [17]).

A study carried out by Kambo and Dutta [18] focused on the applications of hydrochars, depending on their chemical composition, morphological features, and surficial functionalities. Applications can be energy production by combustion, carbon sequestration and gas adsorbent, soil amelioration and activated carbon, for example, for air and water cleaning.

In this study, the feedstock processed was brewer's spent grains (BSG) (dry mass 20–25 wt.%), which is a by-product of breweries; it is lignocellulose with glucamine-rich proteins [19]. The high protein content of BSG makes it suitable for animal nutrition (e.g., cattle), although the proteins cannot be completely assimilated into rumens [20]. Furthermore, the production of BSG is not constant throughout the year; harvesting cycles can result in overproduction. The presence of proteins reduces storability by enzymatic hydrolysis, which is a serious environmental problem. Therefore, there is a real necessity to develop flexible processes that can buffer the BSG quantity. In this context, the implementation of thermochemical processes such as HTC are considered a promising alternative [21].

Regarding this objective, this study tries to contribute to the knowledge of the HTC process under different operating conditions, in order to control the product distribution, maximize the energy yield and evaluate the effect on the main chemical compounds in liquid phase. The final aspect is important in view of the water treatment technologies needed.

The process was carried out in a batch reactor, where the influence of temperature, residence time, and moisture content was evaluated. A two-steps HTC process and the addition of an organic acid (CH_3COOH) is also investigated. The two first parameters, temperature and reaction time, have been considered by several authors [7,22–24], which concluded that product distribution was mainly affected by operating temperature and that the reaction time does not play a significant role. Usually the moisture content was not evaluated. Thus, in this paper we studied two different known moisture contents, three different temperatures, and two reaction times. The different water contents may influence the hydrolysis or polymerization, in the case of limited solubility.

In order to improve the properties of the hydrochars obtained, the strategy of carrying out the process in two-steps was studied, which was considered a significant factor according to Fakkaew et al. [7]. This strategy consists of a first step to enhance the hydrolysis reaction between 170 °C and 180 °C, and a second step, in which the carbonization reactions take place at a range of 200–220 °C. This implies the maximization of both reaction rates, which have different optimums, leading to an increase of fixed carbon (FC) content. On the other hand, Ghanim et al. [25] studied the effect of addition of H_2SO_4 and CH_3COOH. Both had an effect on the removal of ash content; however, H_2SO_4 also produced an increase of FC content. Obtaining a lower ash content because of the removal of alkaline and alkaline earth metals (AAEMs) [26] improves the high heating value (HHV) (MJ/kg) of the hydrochars obtained.

From these prior results the hypothesis was created that two-steps carbonization with a hydrolysis step and the addition of acids improves hydrochar properties as a fuel. This could avoid the increase of temperature to get a similar improvement. If hydrolysis is important, the water content may also have an influence. The aim of this paper is to verify or refute the hypothesis and investigate the influence of water on the two-steps process.

2. Materials and Methods

2.1. Feedstock

The biomass feedstock used in this study is BSG, with a moisture content of 78 wt.%, from Hoepfner Brewery Factory (Karlsruhe, Germany). It was stored at −15°C until processed. In Table 1 (see method section), the main analytical data to characterize the feedstock are summarized.

Table 1. Characterization of brewer's spent grains (BSG). db: dry basis.

Parameters	Units	Values
Dry solid content (105 °C)	wt.%	3.75
Volatile matter (VM, 950 °C)	wt.% (db)	76.25
Ash content (750 °C)	wt.% (db)	4.01
Fixed carbon (FC) content	wt.% (db)	16.00
Higher heating value (*HHV*)	MJ kg^{-1} (db)	22.25
C	wt.% (db)	51.27
H	wt.% (db)	6.97
N	wt.% (db)	4.68
S	wt.% (db)	0.29
Hemicellulose content	wt.% (db)	43.03
Cellulose content	wt.% (db)	23.69
Lignin content	wt.% (db)	5.78
Extractives	wt.% (db)	15.56
Proteins	wt.% (db) [1]	29.26

[1] Standard method (American Society for Testing and Materials (ASTM) D-5291 [27]) was used to calculate protein fraction for BSG on a dry basis by multiplying the total nitrogen value by a factor of 6.25.

2.2. Experimental Procedure

HTC experiments were conducted in an autoclave reactor (VA2 stainless steel) with a volume of 250 mL. To check reproducibility, experiments were performed twice. The feedstock moisture content selected varied between 80 wt.% and 90 wt.% and the initial moisture was adjusted with deionized water. The blend of distilled water and feedstock was stirred manually in order to obtain a homogenous slurry. Then, the reactor was closed and heated inside a gas-chromatography (GC) oven (Hewlett Packard, GC 5890, Koblenz, Germany) (Figure A1). Three different temperatures (180, 200, and 220 °C) were selected and two reaction times (time applied after preheating time) were established: 2 and 4 h. It is necessary to point out that 1 h is required to reach the desired temperature inside the reactor. During the experiments, pressure was measured with a digital pressure gauge and the temperature with a thermocouple. Both were recorded during all processes with a portable data logger (Endress + Hauser, RSG 30, Nesselwang, Germany). Once the experiment had taken placed, the reactor was cooled down to room temperature in 30 min with a cold water bucket.

Afterwards, the total gas volume was determined with the water displacement by gas from a probe full of water. The slurry product from HTC was filtered with a quantitative filter grade 413 VWR® filter paper (VWR European Cat, Leuven, Germany) placed onto a Buchner settle in a flask bottle and connected to a vacuum pump. Liquid and solid phases were weighed to calculate yields of different phases. To remove moisture, the solid phase was dried inside an oven over 24 h at 105 °C up to stable weight. The liquid phase pH was measured by a HACH HQ40d multi equipment and kept in the fridge at 4 °C for further analysis. The solid samples (raw material or hydrochars) were ground into a range of sizes between 150 μm and 250 μm to facilitate its characterization.

In order to obtain hydrochars with the highest *HHV* (MJ/kg), which are related with low O/C and H/C ratios, a two-steps procedure was applied. Two-steps temperature selection was based on the fact that hydrolysis reactions in the HTC process occur below 180 °C and carbonization reactions are favored at 220 °C [7,12]. The first step is a reaction at 180 °C for 1 h. Secondly, a 3 h reaction time, including the time necessary to achieve 220 °C, was applied.

As acids catalyzed the hydrolysis reaction and remove the ash [11], the catalytic effect of acetic acid in the HTC process was also evaluated. For this objective, experiments with 5 wt.% of acetic acid were performed.

2.3. Characterization of Biomass and Products

2.3.1. Solid Fraction

- Proximate analysis

Moisture content, volatile matter (VM) and ash content were determined according to the standards analysis method ASTM D1762-84 [28].

- Elemental analysis

An automatic elemental analyzer EuroEA 3000 Serie (EuroVector S.P.A, Milano, Italy), CHNS-O equipped with a thermal conductivity detector (TCD) was used to determine the content of C, H, N and S of initial biomass and solid products (hydrochars). Samples had already been dried in an oven over 24 h at 105 °C to remove moisture content.

- Fiber analysis

Fiber analysis was performed by Fibretherm FT12 (Königswinter, Germany). Samples were dried at 105 °C for 24 h and cooled in a desiccator before analysis. The cellulose, hemicellulose, lignin and water extractives content of the solid sample was based on Van Soest's method to calculate neutral detergent fiber (NDF), acid detergent fiber (ADF), and acid detergent lignin (ADL) [29].

- Thermogravimetric analysis (TGA)

Hydrochars were ground and sieved to have a particle size smaller than 150–200 µm. The thermal behavior of hydrochars was analyzed with a Netzsch STA Jupiter 449 F5 thermogravimetric balance (Ahlden, Germany). The sample size was approximately 20 mg, heated from room temperature to 105 °C with a heating rate of 10 K min^{-1} over 10 min to remove possible moisture from the sample. After this pretreatment step, the sample was heated to 800 °C with a constant heating rate of 10 K min^{-1} under a constant nitrogen flow of 70 mL min^{-1}.

2.3.2. Liquid Fraction

In order to have a better understanding of changes which occur in the liquid phase after HTC, the total organic carbon (TOC) was measured along with the concentration of chemical compounds through high performance liquid chromatography (HPLC).

- TOC

The TOC of liquid samples was determined using a TOC Analyzer 5050A (Shimadzu Scientific Instruments, Columbia, MD, USA).

- HPLC

The liquid fraction of HTC includes high value molecules (lactic acid, formic acid, acetic acid, levulinic acid (LA), propionic acid, 5-hydroxymethylfurfural (HMF), and furfural), which were determined using HPLC (Shimadzu 20AD, Shimadzu, Canby, OR, USA). This is equipped with a column suitable for organic acids, called Aminex HPX-87H, UV-vis detector (SPD-20A, Shimadzu), and refractive index detector (RID-10A, Shimadzu). The mobile phase consists of a 4 mM solution of H_2SO_4 in water and a flow rate of 0.6 mL min^{-1}. The column temperature was set at 35 °C. The time required for a complete analysis of each sample was 60 min. Filtration was required prior to HPLC analysis with a 0.2 µm PFTE (VWR International, Radnor, PA, USA). A 10 µL volume of each sample was injected to determine the concentration of chemical compounds.

2.3.3. Fuel Analysis

From Table 2 where 80 and 90 wt.% are the initial moisture contents; 180, 200, and 220 °C the temperatures; and 2 and 4 h the reaction times—the fuel analysis was done using the Equations (1)–(5). The parameters necessary to describe the energy content of hydrochars are higher heating value (*HHV*) [30], fuel ratio, hydrochar yield (*Hy*) and energy densification (*Ed*). Energy yield (*Ey*) was calculated using the following equations:

$$HHV\left(\frac{\text{MJ}}{\text{kg}}\right) = 0.3491 \times \text{C} + 1.1783 \times \text{H} + 0.1005 \times \text{S} + 0.1034 \times \text{O} + 0.0151 \times \text{N} + 0.0211 \times \text{Ash} \tag{1}$$

$$\text{Fuel ratio} = \frac{\text{FC}}{\text{VM}} \tag{2}$$

$$Hy\,(\%) = \frac{\text{mass of dried hydrochars}}{\text{mass of total dried feedstock}} \times 100 \tag{3}$$

$$Ed = \frac{HHV \text{ of dried hydrochars}}{HHV \text{ of dried feedstock}} \tag{4}$$

$$Ey\,(\%) = \text{Hydrochar yield} \times Ed \tag{5}$$

Table 2. Proximate analysis and fuel (see methods for description).

Samples	Proximate Analysis (wt.% Dry Basis)			Fuel Ratio	*Hy* (%)	*Ey* (%)	*Ed*
	VM	Ash	FC				
Raw material	77.87	4.30	17.83	0.23	-	-	-
HTC-80-180-2	74.92	4.29	20.79	0.28	68.00	80.81	1.19
HTC-80-200-2	71.96	4.29	23.75	0.33	64.07	78.09	1.22
HTC-80-220-2	69.03	4.26	26.71	0.39	57.99	76.10	1.31
HTC-80-180-4	72.79	4.18	23.03	0.32	67.52	83.54	1.24
HTC-80-200-4	70.50	4.15	25.35	0.36	63.48	76.60	1.21
HTC-80-220-4	66.17	4.22	29.61	0.45	55.04	72.43	1.32
HTC-90-180-2	74.22	3.39	22.39	0.30	66.17	79.49	1.20
HTC-90-200-2	71.69	3.24	25.07	0.35	62.12	76.84	1.24
HTC-90-220-2	68.04	3.22	28.74	0.42	52.04	69.14	1.33
HTC-90-180-4	73.15	3.12	23.73	0.32	65.26	76.79	1.18
HTC-90-200-4	70.33	3.16	26.50	0.38	60.53	76.21	1.26
HTC-90-220-4	64.86	3.27	31.88	0.49	50.59	66.02	1.31
HTC-80-180/220-4	65.84	4.11	30.05	0.46	67.05	85.51	1.28
HTC-90-180/220-4	65.09	3.12	31.79	0.49	70.30	90.64	1.29
HTC-80-220-acid	61.54	3.67	34.79	0.57	60.36	79.11	1.31
HTC-90-220-acid	59.35	2.81	37.84	0.64	52.56	71.28	1.36

3. Results and Discussion

The results shown in Tables 1–4 are expressed as the average of values obtained with a standard deviation below 5%. Due to the calibration of the HPLC the results obtained (Table 5) are the average of the values with a standard deviation below 7%.

3.1. Carbon Balance of Brewer's Spent Grains during Hydrothermal Carbonization

The characteristics of native BSG were calculated according to Section 2.3.1. The values obtained are comparable to data reported previously [31,32]. The C content was measured in both solid and liquid phase; it cannot be measured in gas phase, because the volume was too small.

Equations (6)–(11) were used to have a better overview of the carbon distribution in feedstock and products:

Mass of carbon in the feedstock solid phase:

$$m(\text{carbon})_{\text{feedstock solid}} = m_{\text{feedstock solid}} \times x(\text{carbon})_{\text{feedstock}} \tag{6}$$

$$m(\text{carbon})_{\text{feedstock liquid}} = m_{\text{feedstock liquid}} \times \frac{\text{TOC}}{\rho} \quad (7)$$

Products equations for the mass of carbon in the solid and liquid phase, respectively:

$$m(\text{carbon})_{\text{solid}} = m_{\text{solid}} \times x(\text{carbon})_{\text{solid}} \quad (8)$$

$$m(\text{carbon})_{\text{liquid}} = m_{\text{liquid}} \times \frac{\text{TOC}}{\rho} \quad (9)$$

Distribution of C into products:

$$\%\text{C}_{\text{solid}} = \frac{m(\text{carbon})_{\text{solid}}}{\left(m(\text{carbon})_{\text{solid}} + m(\text{carbon})_{\text{liquid}}\right)} \quad (10)$$

$$\%\text{C}_{\text{liquid}} = \frac{m(\text{carbon})_{\text{liquid}}}{\left(m(\text{carbon})_{\text{solid}} + m(\text{carbon})_{\text{liquid}}\right)} \quad (11)$$

The use of Equations (6)–(11) and experimental results obtained are showed in Table 3. It can be seen that at the same moisture content but higher temperature, % C content changes and less carbon is in the solid phase (Figure 1). However, with a 90% moisture content and 4 h reaction time, it is observed that % C content remains temperature independent in the solid phase (Figure 1). In general, the increase of the % C content (Figure 1) and the increase of carbon in solid phase is possible due to the polymerization reactions of the monomers solved of the liquid phase, which are formed during the hydrolysis [12]. This influence of the hydrolysis reaction was confirmed with a two-steps reaction, because of the increment of C on the solid phase (90%, 220 °C, 4 h). A first step, with 1 h at a low temperature (180 °C), enhances hydrolysis reactions and leads to the rupture of the initial poly-sugars into monomers, which remain in liquid phase [33]. During the second step, consisting of 3 h reaction time at 220 °C, water elimination to furfural-rings occurs. This loss of water is the "carbonization", because the content of carbon increases accordingly. These intermediates polymerize and increase the content of C in the solid phase. However, a certain amount of C remains into liquid phase (Figure 1) [34].

Table 3. Distribution of C into solid and liquid phase.

Samples	Solid wt.%	Liquid wt.%
HTC-80-180-2	0.86	0.14
HTC-80-200-2	0.86	0.14
HTC-80-220-2	0.84	0.16
HTC-80-180-4	0.86	0.14
HTC-80-200-4	0.84	0.16
HTC-80-220-4	0.84	0.16
HTC-90-180-2	0.83	0.17
HTC-90-200-2	0.82	0.18
HTC-90-220-2	0.78	0.22
HTC-90-180-4	0.82	0.18
HTC-90-200-4	0.81	0.19
HTC-90-220-4	0.82	0.18
HTC-80-180/220-4	0.85	0.15
HTC-90-180/220-4	0.84	0.16
HTC-80-220-acid	0.71	0.29
HTC-90-220-acid	0.66	0.34

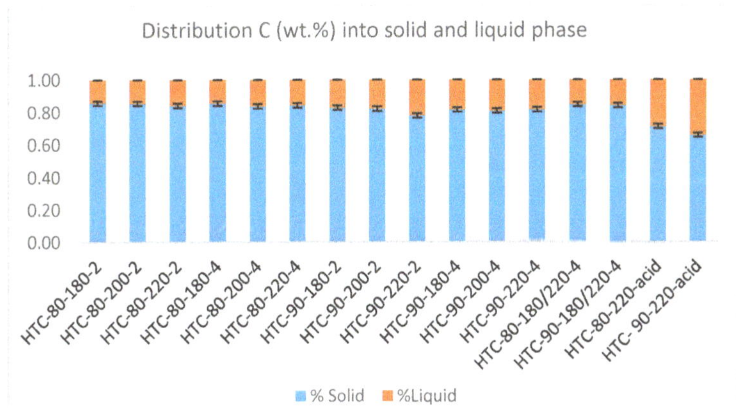

Figure 1. Distribution of C into solid, liquid and gas phase.

3.2. Characteristics of Hydrochars

Proximate analysis of hydrochars produced under different reaction conditions are showed in Table 2. The VM content decreases with an increase of temperature and reaction time, with different possible explanations. One is the enhancement of dehydration and decarboxylation reactions, increasing the carbon content in the solid from more coal-like material. With increasing severity, the number of crosslinking reactions increases, leading less low-molecular and therefore less volatile molecules. On the other hand, less carbon is found in solid phase (Table 2), which could imply the conversion of VM during HTC into solved compounds in the liquid phase. Both significantly influence the ignition behavior [35,36]. The comparison between VM content under the same conditions, but with different feedstock moisture content, showed no significantly higher value with 80% than with 90% moisture content. In contrast, the FC content of hydrochars increased with temperature and reaction time, in accord with findings of previous studies [37].

The hydrochar ash content was independent of temperature and reaction time. However, it decreased with higher moisture content because the same quantity of ash was in contact with a higher quantity of water. The increase of ash content at 220 °C and a 4 h reaction time with moisture content can be explained by the re-precipitation of some inorganic components [38].

To understand the characteristics of the hydrochars produced, it is necessary to understand the fuel ratio (Equation (2)). The hydrochars produced at 220 °C, with a 4 h reaction time and moisture content of 80% and 90% contains the highest FC, with 29.61% and 31.88%, respectively, and contains the lowest VM of 66.17% (at 80% moisture content) and 64.84% (at 90% moisture content). Consequently, the fuel ratios obtained under these conditions are twice as large as those of the initial feedstock (Table 2).

The yield to hydrochars (*Hy* %) at low reaction temperature has the highest values; the low *HHV* (MJ/kg) [39] of this hydrochar causes the *Ey* of all samples to have similar values (Table 2). This is reported in the literatures [40,41].

The values of Ed obtained were over 1, showing the improvement of energy densification by HTC [42]. As results using varied moisture contents were similar, a value of 90% moisture content was chosen to carry out experiments with the two-steps process and varied acid addition. In comparison with experiments at 180 °C, 4 h and 90% moisture content, performing the reaction in two-steps and incorporating 5% wt. of acid had the effect of reducing the VM content to similar values reached at 220 °C, 4 h and 90% moisture content (Table 3). The two-steps reaction process leads to energy saving, because it produced the same reduction of VM with less energy consumption to carry out the HTC process. Acid addition shows a significant importance, as shown by Ghanim et al. [25], wherein the addition of acetic acid not only reduced ash (%), it also produced an increment of FC (%), which implied an increment of fuel ratio and *Ey* (%). The reason for this is that polymerization, water elimination and decarboxylation (all reactions) increase the heating value [34,43].

Table 4 summarizes the elemental composition of hydrochars obtained under different operation conditions. According to Funke et al. [12], the atomic ratios of H/C and O/C decreased due to chemical defunctionalization (dehydration reaction and decarboxylation). This relationship between O/C and H/C was plotted as a Van Krevelen diagram (Figure 2). The higher ratios of O/C and H/C, which belong to the original biomass, are plotted in the upper right corner of the diagram. The increment of the temperature and reaction time in HTC reduces the ratios until the typical area of lignite is reached, which is similar to results of others authors [2,18]. The acidification of initial feedstock produced the lowest ratio of O/C and H/C, which can be explained by the reduction of hydroxyl groups by water elimination, which also increased the hydrophobicity of the hydrochar [29]. The reduction of the O/C ratio to half that of the initial biomass is due to the high content of hemicellulose; hemicellulose is the most reactive part of the feedstock biomass. A study carried out by Wikberg et al. [44] showed that a decrement of O/C ratio in coffee cake was due to removal of carboxyl groups from biomass extractives, hemicellulose, and cellulose. Therefore, it can be assumed that oxygen and hydrogen of the initial biomass migrates as water, mainly to the liquid phase [45].

Table 4. Elemental analysis, O/C and H/C ratios and high heating values (*HHV*).

Samples	Elemental Analysis (% wt. Dry Basis)					O/C	H/C	*HHV* (MJ/kg)
	N	C	H	S	O			
Raw material	4.68	51.27	6.97	0.29	36.52	0.53	1.63	22.28
HTC-80-180-2	4.37	60.32	7.07	0.45	27.80	0.35	1.41	26.48
HTC-80-200-2	4.12	62.02	7.02	0.42	26.42	0.32	1.36	27.16
HTC-80-220-2	4.41	65.81	7.28	0.43	22.07	0.25	1.33	29.24
HTC-80-180-4	4.30	62.89	7.02	0.43	25.36	0.30	1.34	27.57
HTC-80-200-4	4.09	61.19	7.09	0.46	27.17	0.33	1.39	26.89
HTC-80-220-4	4.46	67.08	6.89	0.43	21.14	0.24	1.23	29.32
HTC-90-180-2	3.63	60.18	7.39	0.44	28.36	0.35	1.47	26.76
HTC-90-200-2	3.80	62.29	7.25	0.45	26.21	0.32	1.40	27.56
HTC-90-220-2	3.75	66.55	7.35	0.41	21.94	0.25	1.33	29.60
HTC-90-180-4	3.49	59.96	7.05	0.42	29.07	0.36	1.41	26.22
HTC-90-200-4	3.86	63.54	7.19	0.45	24.96	0.29	1.36	28.05
HTC-90-220-4	4.25	66.60	6.89	0.43	21.83	0.25	1.24	29.08
HTC-80-180/220-4	4.40	64.84	6.97	0.49	23.30	0.27	1.29	28.42
HTC-90-80/220-4	4.37	65.46	6.99	0.53	22.66	0.26	1.28	28.73
HTC-80-220-acid	4.55	67.29	6.71	0.51	20.94	0.23	1.20	29.20
HTC-90-220-acid	4.20	68.67	7.03	0.50	19.60	0.21	1.23	30.22

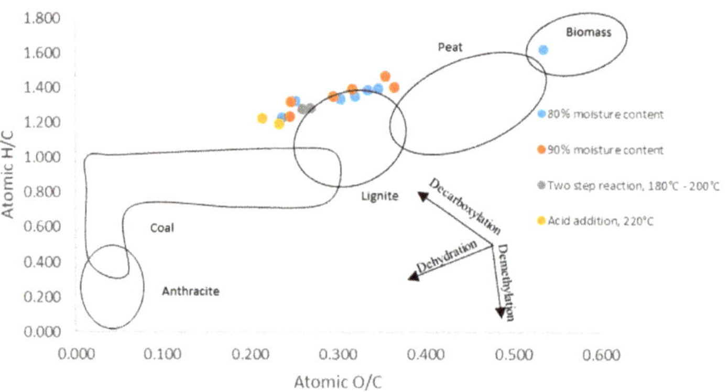

Figure 2. Van Krevelen diagram for hydrochars versus temperature, residence time and moisture content.

On the other hand, S (%) was constant across the HTC condition range and N (%) decreased with higher moisture content of feedstock and higher temperatures. The high protein and hemicellulose content of BSG (Table 1) has an effect during hydrolysis. Amino acids and amines, as consecutive products of the hydrolysis of proteins, can react with carbonyl groups. This Maillard reaction starts at 180 °C and leads, for example, to *N*-heterocycles [46–48]. These heterocycles become part of the hydrochar. With a higher water content, more *N*-compounds are solved and not incorporated in the hydrochar. In addition, at higher temperatures, *N*-containing functional groups are hydrolyzed [16].

3.3. Thermogravimetric Analysis

The TGA of hydrochars was done for 90% moisture content (regarding feedstock) samples and 4 h of reaction time. The TGA (Figure 3) of hydrochars at 180 °C and 220 °C are used to compare hydrochars produced by the one-step procedure with the hydrochars obtained with two-steps reaction and acid addition. Figure 3 shows the derivative mass loss (DTG) for BSG and hydrochars produced at 180 °C, 220 °C, and in the two-steps reaction between 180 °C and 220 °C and with acid addition at 220 °C. The DTG curves show four characteristic peaks at 251, 287, 350, and 421 °C. The peaks correspond to the three biopolymers of hemicellulose, cellulose and lignin, which built the biomass structure. The composition of the origin biomass is shown in Table 1 and consists of hemicellulose (43.03%), cellulose (23.69%), and lignin (5.78%). However, it was mentioned in Section 2.3.1 that samples were kept at 105 °C for 10 min, so moisture peaks and volatile peaks shown in the reviewed literatures [31,37] at a range of 80 °C to 105 °C, did not appear.

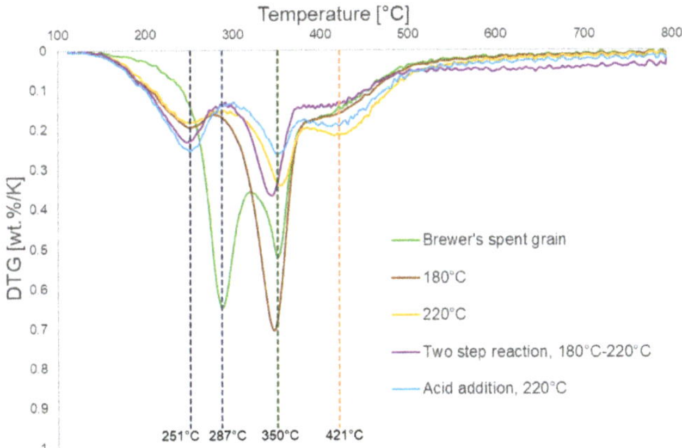

Figure 3. Hydrochar thermal stability at 90% moisture content during 4 h reaction time.

Raw feedstock has a largest peak at 287 °C, which corresponds to hemicellulose decomposition, because hemicellulose is an amorphous polymer and therefore less thermally stable [49]. In addition, it makes it more feasible for hydrolysis reactions than the more crystalline cellulose (see below). Cellulose decomposition has a peak similar to hemicellulose but at a higher temperature [50].

DTG curves of the different hydrochars with the same reaction time but different temperatures (Figure 3) are similar to those of previous studies [14,51]. The peak of hemicellulose disappearing agrees with research published by Kruse et al. [51], which explains that the HTC process leads to degradation of hemicellulose at 180 °C. However, there is a shoulder at 251 °C as a result of the decomposition of feedstock during HTC into more reactive compounds than hemicellulose.

In comparison to the DTG of BSG, hydrochars have a small peak at temperatures around 420 °C. This peak is larger at a higher hydrochar formation temperature. Therefore, it can be attributed to carbonization during HTC leading to short-chained polymers, which are less stable in TGA than the original biomass [51]. It has to be stated here that polycarbohydrates like hemicellulose, and especially cellulose, are stabilized by intermolecular H-bonds, which is not possible in hydrochars. In hydrochars the necessary OH-groups are missing because of water elimination during HTC. Consequently, smaller molecules of hydrochar are evaporated at lower temperatures than carbohydrates fixed in a structure by H-bonds.

The effect of a two-steps reaction and acid addition during HTC on TGA analysis (Figure 3) shows that both improve the hydrolysis of the initial feedstock. This means that fewer residual polycarbohydrates and more carbonization products are found. Both treatments produce more thermal volatile compounds because their peak at a temperature of 251 °C is more than the peak for hydrochars without acid addition or application of the two-steps reaction. Acid addition decreases the peak of cellulose due to catalytic effect of acids on the hydrolysis of cellulose during the HTC process.

In summary, two effects influence the thermal properties of hydrochar. On one hand, the temperature increase results in a decrease of the VM content in the hydrochars (Table 2), producing a thermally stable material with higher *HHV* (MJ/kg). This is a consequence of stronger crosslinking and the formation of bonds. On the other hand, less thermal stable compounds are formed. For hydrochars there occurs a DTG peak at 251 °C, which may be related with degradation of parts of hydrochars with less stable structures, as well as evaporation of organic substances (HMF), created during HTC. The last DTG peak at 421 °C is related with lignin decomposition [52]. In addition, this peak might be also caused by hydrochar degradation. For hydrochars produced at 180 °C and 220 °C, the peak is increased. This may be due to the decomposition of less stable intermediate carbonization products.

3.4. Characteristics of Liquid Phase

The pH values of different HTC liquid phases are shown in Table 5. The liquid phase after HTC at a temperature of 180 °C has a pH of around 4 and increases to pH 4.6 with increasing reaction temperature and reaction time. It supports the idea that lower temperatures promote hydrolysis reactions, eliciting the release of organic acids, such as acetic acid. Afterwards, these organic acids show further reactions at higher temperatures [53,54]. This supports the formation of other compounds, because hydrolysis is catalyzed by acids. Therefore, the two-steps experiments and acid addition were performed to support the idea of hydrolysis and consecutive re-polymerization mechanisms. Two-steps experiments accomplished during this work showed that 1 h at a temperature of 180 °C is not enough to enhance the promotion of organic acids, because the pH of the liquid phase is slightly lower than for the experiments accomplished at 220 °C, during 4 h of reaction time. At a moisture content of 80% the pH value is 4.53, slightly higher than without acid addition (pH 3.57). The influence of pH level is shown by the yield to organic chemical molecules in the liquid fraction, mainly lactic acid, formic acid, acetic acid, levulinic acid (LA), and propionic acid (Table 5) [55]. These acids are usually found in hydrothermal conversions. Here they are relatively inert molecules, because they have no, or a low, tendency to polymerize. In addition, the yield of the organic chemical compounds in the liquid phase is related with the TOC. The highest values of TOC were obtained at low temperatures, where hydrolysis reactions were promoted but the formation of a solid product by polymerization is too slow.

Figure 4 shows the yields of the main low molecular weight acids (lactic, formic, acetic, levulinic, and propionic acid) produced during HTC. Lactic acid can be obtained through conversion of carbohydrates [56,57] under HTC conditions [57]. The yield trends to increase with the temperature from 180 °C to 220 °C. This fact could be to the production of trioses (glycolaldehyde, diehydroxyacetone and other tautomers) from biomass, which are intermediates produced during retro-aldol condensation of sugars [58]. The trioses further react to lactic acids. According to Zan et al. [59], the formation of lactic acid is not affected by the presence of formic acids; however, in this study (Figure 4a) reflect that higher presence of formic acid is correlated with a lower yields of lactic acid. At low moisture content and 200 °C the highest yield to acetic and propionic acids were found (Figure 4c,e). It is supposed that acetic acid was produced due to oxidation reaction of acetaldehyde. Acetaldehyde was produced via decarbonylation reactions of lactic acid [60]. Lactic acid is also supposed to be converted by dehydration reaction to acrylic acid. Afterwards, hydrogenation of acrylic acid produce propionic acid [60]. In sum, these sequences of reactions are very speculative. Only the primary products can be easily identified.

Table 5. pH and dissolved organic chemical yield in liquid fraction at different process conditions. HMF: 5-hydroxymethylfurfural and LA: levulinic acid.

Samples	pH	Lactic Acid	Formic Acid	Acetic Acid	Propionic Acid	HMF	Furfural	LA
HTC-80-180-2	4.00	0.06	0.09	0.10	0.03	0.01	0.01	0.01
HTC-80-200-2	4.38	0.04	0.08	0.11	0.02	0.00	0.00	0.01
HTC-80-220-2	4.53	0.10	0.05	0.09	0.01	0.00	0.00	0.01
HTC-80-180-4	4.16	0.04	0.08	0.10	0.02	0.00	0.00	0.01
HTC-80-200-4	4.53	0.10	0.05	0.11	0.00	0.00	0.00	0.01
HTC-80-220-4	4.53	0.08	0.04	0.10	0.00	0.00	0.00	0.01
HTC-90-180-2	3.91	0.02	0.06	0.05	0.03	0.01	0.02	0.00
HTC-90-200-2	4.42	0.05	0.08	0.08	0.08	0.00	0.00	0.01
HTC-90-220-2	4.56	0.06	0.06	0.07	0.05	0.00	0.00	0.01
HTC-90-180-4	4.04	0.04	0.08	0.07	0.02	0.00	0.00	0.01
HTC-90-200-4	4.59	0.04	0.01	0.05	0.01	0.00	0.00	0.00
HTC-90-220-4	4.66	0.06	0.04	0.10	0.00	0.00	0.00	0.01
HTC-80-180/220-4	4.64	0.08	0.04	0.10	0.10	0.00	0.00	0.01
HTC-90-180/220-4	4.54	0.09	0.05	0.09	0.17	0.00	0.00	0.01
HTC-80-220-acid	3.67	0.07	0.03	106.40	0.15	0.00	0.00	0.02
HTC-90-220-acid	3.57	0.05	0.03	100.52	0.11	0.00	0.00	0.02

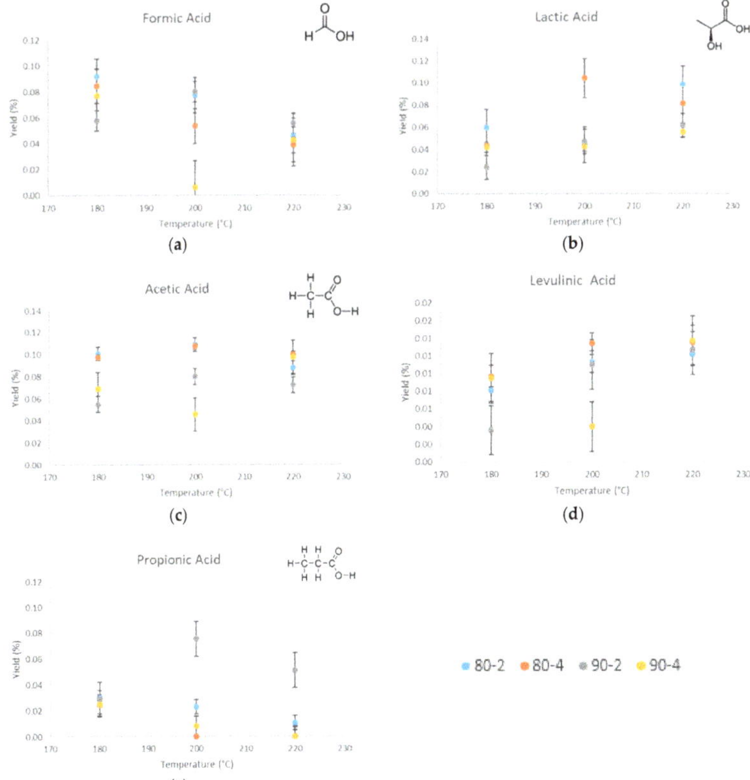

Figure 4. Yield of (**a**) formic acid, (**b**) lactic acid, (**c**) acetic acid, (**d**) LA, and (**e**) propionic acid at experimental conditions.

In addition, the production of levulinic acid (LA) (Figure 4d), which is one of the products of the rehydration of HMF in aqueous media, is observed [61]. Table 5 shows that the addition of acid into the initial slurry produces higher amounts of levulinic acid. According to Licursi et al. [62], it can be due to the catalytic effect of acetic acid on the pH-dependent hydrolysis reaction of cellulose in the initial biomass, which started at temperatures over 200 °C. However, it is more likely that it is because HMF and levulinic acid formation is slightly different. Therefore, the addition of an acid changes the selectivity to LA. LA is preferred at higher acid concentrations due to the selection of LA increasing during the HMF decomposition into LA and humins [43].

4. Conclusions

In this study, we examined the behavior of BSG during HTC and the partition of C into solid and liquid phase under different reaction conditions. These parameters are temperature and reaction time, as well as moisture content. In the experimental range investigated here, the HTC process itself is independent of moisture content, that is a new feedstock variable. The ash content decreases in the case with higher water content. Acid addition produced an increase of carbon distribution to the liquid. The highest fuel ratio of hydrochars was produced at 220 °C, with a 4 h reaction time and acid addition, and TGA curves showed the complete hydrolysis of the hemicellulose and cellulose. Thus, this hydrochar can be used as an ecofriendly solid fuel.

Hydrochar products of a two-steps reaction have higher values to *Hy* (%) than those produced in a single-step process. The fuel properties of solids are also better than with a single-step reaction. Therefore, two-steps reactions have an economic benefit and lead to an improvement of solid product characteristics.

The HPLC analysis of the different HTC conditions shows different small acids. These are consecutive products of the carbohydrate splitting reactions. They are found because they do not polymerize to hydrochar, or react only with a very low reaction rate.

Author Contributions: P.J.A. performed the investigation and wrote the article draft, M.P.O. performed and discussed the TGA results, and A.K. reviewed the article.

Funding: This research was funded by the European Union's Horizon 2020 research and innovation program under the Marie Skłodowska-Curie Grant Agreement No. 721991.

Acknowledgments: We gratefully acknowledge the work of doctoral candidate Paul Körner (HPLC).

Conflicts of Interest: The authors declare no conflict of interest.

Appendix A

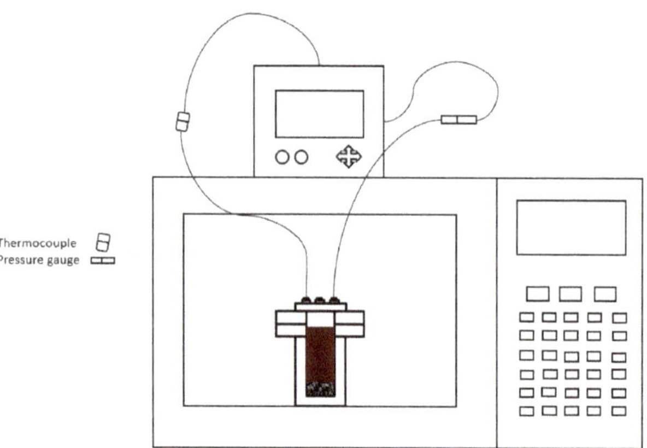

Figure A1. Experimental set up of HTC into gas-chromatography (GC) oven.

References

1. Liu, Z.; Quek, A.; Kent Hoekman, S.; Balasubramanian, R. Production of solid biochar fuel from waste biomass by hydrothermal carbonization. *Fuel* **2013**, *103*, 943–949. [CrossRef]
2. Basso, D.; Patuzzi, F.; Castello, D.; Baratieri, M.; Rada, E.C.; Weiss-Hortala, E.; Fiori, L. Agro-industrial waste to solid biofuel through hydrothermal carbonization. *Waste Manag.* **2016**, *47*, 114–121. [CrossRef] [PubMed]
3. Peterson, A.A.; Vogel, F.; Lachance, R.P.; Fröling, M.; Antal, M.J., Jr.; Tester, J.W. Thermochemical biofuel production in hydrothermal media: A review of sub- and supercritical water technologies. *Energy Environ. Sci.* **2008**, *1*, 32–65. [CrossRef]
4. Libra, J.A.; Ro, K.S.; Kammann, C.; Funke, A.; Berge, N.D.; Neubauer, Y.; Titirici, M.-M.; Fühner, C.; Bens, O.; Kern, J.; et al. Hydrothermal carbonization of biomass residuals: A comparative review of the chemistry, processes and applications of wet and dry pyrolysis. *Biofuels* **2014**, *2*, 71–106. [CrossRef]
5. Gao, P.; Zhou, Y.; Meng, F.; Zhang, Y.; Liu, Z.; Zhang, W.; Xue, G. Preparation and characterization of hydrochar from waste eucalyptus bark by hydrothermal carbonization. *Energy* **2016**, *97*, 238–245. [CrossRef]
6. Guizani, C.; Jeguirim, M.; Valin, S.; Limousy, L.; Salvador, S. Biomass Chars: The Effects of Pyrolysis Conditions on Their Morphology, Structure, Chemical Properties and Reactivity. *Energies* **2017**, *10*, 796. [CrossRef]
7. Fakkaew, K.; Koottatep, T.; Polprasert, C. Effects of hydrolysis and carbonization reactions on hydrochar production. *Bioresour. Technol.* **2015**, *192*, 328–334. [CrossRef] [PubMed]
8. Tsukashima, H. The Infrared Spectra of Artificial Coal made from Submerged Wood at Uozu, Toyama Prefecture, Japan. *Bull. Chem. Soc. Jpn.* **1966**, *39*, 460–465. [CrossRef]
9. Mäkelä, M.; Benavente, V.; Fullana, A. Hydrothermal carbonization of lignocellulosic biomass: Effect of process conditions on hydrochar properties. *Appl. Energy* **2015**, *155*, 576–584. [CrossRef]
10. Gollakota, A.R.K.; Kishore, N.; Gu, S. A review on hydrothermal liquefaction of biomass. *Renew. Sustain. Energy Rev.* **2018**, *81*, 1378–1392. [CrossRef]
11. Yanik, J.; Ebale, S.; Kruse, A.; Saglam, M.; Yuskel, M. Biomass gasification in supercritical water: Part 1. Effect of the nature of biomass. *Fuel* **2007**, *86*, 2410–2415. [CrossRef]
12. Funke, A.; Ziegler, F. Hydrothermal carbonization of biomass: A summary and discussion of chemical mechanisms for process engineering. *Biofuels Bioprod. Bioref.* **2010**, *4*, 160–177. [CrossRef]
13. Yoshikawa, K.; Prawisudha, P. Hydrothermal Treatment of Municipal Solid Waste for Producing Solid Fuel. In *Application of Hydrothermal Reactions to Biomass Conversion*; Jin, F., Ed.; Springer: Heidelberg, Germany; New York, NY, USA, 2014; pp. 355–383.
14. Kang, S.; Li, X.; Fan, J.; Chang, J. Characterization of Hydrochars Produced by Hydrothermal Carbonization of Lignin, Cellulose, d-Xylose, and Wood Meal. *Ind. Eng. Chem. Res.* **2011**, *51*, 9023–9031. [CrossRef]
15. Sevilla, M.; Fuertes, A.B. The production of carbon materials by hydrothermal carbonization of cellulose. *Carbon* **2009**, *47*, 2281–2289. [CrossRef]
16. Kruse, A.; Koch, F.; Stelzl, K.; Wüst, D.; Zeller, M. Fate of Nitrogen during Hydrothermal Carbonization. *Energy Fuels* **2016**, *30*, 8037–8042. [CrossRef]
17. Karayıldırım, T.; Sınağ, A.; Kruse, A. Char and Coke Formation as Unwanted Side Reaction of the Hydrothermal Biomass Gasification. *Chem. Eng. Technol.* **2008**, *31*, 1561–1568. [CrossRef]
18. Kambo, H.S.; Dutta, A. A comparative review of biochar and hydrochar in terms of production, physico-chemical properties and applications. *Renew. Sustain. Energy Rev.* **2015**, *45*, 359–378. [CrossRef]
19. Celus, I.; Brijs, K.; Delcour, J.A. Fractionation and characterization of brewers' spent grain protein hydrolysates. *J. Agric. Food Chem.* **2009**, *57*, 5563–5570. [CrossRef] [PubMed]
20. Santos, M.; Jiménez, J.J.; Bartolomé, B.; Gómez-Cordovés, C.; del Nozal, M.J. Variability of brewer's spent grain within a brewery. *Food Chem.* **2003**, *80*, 17–21. [CrossRef]
21. Celus, I.; Brijs, K.; Delcour, J.A. Enzymatic hydrolysis of brewers' spent grain proteins and technofunctional properties of the resulting hydrolysates. *J. Agric. Food Chem.* **2007**, *55*, 8703–8710. [CrossRef] [PubMed]
22. Pruksakit, W.; Patumsawad, S. Hydrothermal Carbonization (HTC) of Sugarcane Stranded: Effect of Operation Condition to Hydrochar Production. *Energy Procedia* **2016**, *100*, 223–226. [CrossRef]
23. Ulbrich, M.; Preßl, D.; Fendt, S.; Gaderer, M.; Spliethoff, H. Impact of HTC reaction conditions on the hydrochar properties and CO_2 gasification properties of spent grains. *Fuel Process. Technol.* **2017**, *167*, 663–669. [CrossRef]

24. Saba, A.; Saha, P.; Reza, M.T. Co-Hydrothermal Carbonization of coal-biomass blend: Influence of temperature on solid fuel properties. *Fuel Process. Technol.* **2017**, *167*, 711–720. [CrossRef]
25. Ghanim, B.M.; Kwapinski, W.; Leahy, J.J. Hydrothermal carbonisation of poultry litter: Effects of initial pH on yields and chemical properties of hydrochars. *Bioresour. Technol.* **2017**, *238*, 78–85. [CrossRef] [PubMed]
26. Pecha, B.; Arauzo, P.; Garcia-Perez, M. Impact of combined acid washing and acid impregnation on the pyrolysis of Douglas fir wood. *J. Anal. Appl. Pyrolysis* **2015**, *114*, 127–137. [CrossRef]
27. D02 Committee. *Test Methods for Instrumental Determination of Carbon, Hydrogen, and Nitrogen in Petroleum Products and Lubricants*; ASTM D-5291; American Society for Testing and Materials (ASTM) International: West Conshohocken, PA, USA, 1996.
28. D07 Committee. *Test Method for Chemical Analysis of Wood Charcoal*; ASTM D1762-84; ASTM International: West Conshohocken, PA, USA, 2013.
29. Reza, M.T.; Uddin, M.H.; Lynam, J.G.; Hoekman, S.K.; Coronella, C.J. Hydrothermal carbonization of loblolly pine: Reaction chemistry and water balance. *Biomass Conv. Bioref.* **2014**, *4*, 311–321. [CrossRef]
30. Channiwala, S.A.; Parikh, P.P. A unified correlation for estimating HHV of solid, liquid and gaseous fuels. *Fuel* **2002**, *81*, 1051–1063. [CrossRef]
31. Poerschmann, J.; Weiner, B.; Wedwitschka, H.; Baskyr, I.; Koehler, R.; Kopinke, F.-D. Characterization of biocoals and dissolved organic matter phases obtained upon hydrothermal carbonization of brewer's spent grain. *Bioresour. Technol.* **2014**, *164*, 162–169. [CrossRef] [PubMed]
32. Vieira, E.F.; da Silva, D.D.; Carmo, H.; Ferreira, I.M. Protective ability against oxidative stress of brewers' spent grain protein hydrolysates. *Food Chem.* **2017**, *228*, 602–609. [CrossRef] [PubMed]
33. Lachos-Perez, D.; Tompsett, G.A.; Guerra, P.; Timko, M.T.; Rostagno, M.A.; Martínez, J.; Forster-Carneiro, T. Sugars and char formation on subcritical water hydrolysis of sugarcane straw. *Bioresour. Technol.* **2017**, *243*, 1069–1077. [CrossRef] [PubMed]
34. Kruse, A.; Funke, A.; Titirici, M.-M. Hydrothermal conversion of biomass to fuels and energetic materials. *Curr. Opin. Chem. Biol.* **2013**, *17*, 515–521. [CrossRef] [PubMed]
35. Yao, Z.; Ma, X.; Lin, Y. Effects of hydrothermal treatment temperature and residence time on characteristics and combustion behaviors of green waste. *Appl. Therm. Eng.* **2016**, *104*, 678–686. [CrossRef]
36. He, C.; Giannis, A.; Wang, J.-Y. Conversion of sewage sludge to clean solid fuel using hydrothermal carbonization: Hydrochar fuel characteristics and combustion behavior. *Appl. Energy* **2013**, *111*, 257–266. [CrossRef]
37. Chen, X.; Ma, X.; Peng, X.; Lin, Y.; Yao, Z. Conversion of sweet potato waste to solid fuel via hydrothermal carbonization. *Bioresour. Technol.* **2018**, *249*, 900–907. [CrossRef] [PubMed]
38. Reza, M.T.; Lynam, J.G.; Uddin, M.H.; Coronella, C.J. Hydrothermal carbonization: Fate of inorganics. *Biomass Bioenergy* **2013**, *49*, 86–94. [CrossRef]
39. Zhao, X.; Becker, G.C.; Faweya, N.; Rodriguez Correa, C.; Yang, S.; Xie, X.; KRUSE, A. Fertilizer and activated carbon production by hydrothermal carbonization of digestate. *Biomass Conv. Bioref.* **2018**, *8*, 423–436. [CrossRef]
40. Volpe, M.; Fiori, L. From olive waste to solid biofuel through hydrothermal carbonisation: The role of temperature and solid load on secondary char formation and hydrochar energy properties. *J. Anal. Appl. Pyrolysis* **2017**, *124*, 63–72. [CrossRef]
41. Jain, A.; Balasubramanian, R.; Srinivasan, M.P. Hydrothermal conversion of biomass waste to activated carbon with high porosity: A review. *Chem. Eng. J.* **2016**, *283*, 789–805. [CrossRef]
42. Oh, S.-Y.; Yoon, Y.-M. Energy Recovery Efficiency of Poultry Slaughterhouse Sludge Cake by Hydrothermal Carbonization. *Energies* **2017**, *10*, 1876. [CrossRef]
43. Körner, P.; Jung, D.; Kruse, A. The effect of different Brønsted acids on the hydrothermal conversion of fructose to HMF. *Green Chem.* **2018**, *20*, 2231–2241. [CrossRef]
44. Wikberg, H.; Grönqvist, S.; Niemi, P.; Mikkelson, A.; Siika-Aho, M.; Kanerva, H.; Käsper, A.; Tamminen, T. Hydrothermal treatment followed by enzymatic hydrolysis and hydrothermal carbonization as means to valorise agro- and forest-based biomass residues. *Bioresour. Technol.* **2017**, *235*, 70–78. [CrossRef] [PubMed]
45. Licursi, D.; Antonetti, C.; Mattonai, M.; Pérez-Armada, L.; Rivas, S.; Ribechini, E.; Raspolli Galletti, A.M. Multi-valorisation of giant reed (Arundo Donax L.) to give levulinic acid and valuable phenolic antioxidants. *Ind. Crops Prod.* **2018**, *112*, 6–17. [CrossRef]

46. Brunner, G. Near critical and supercritical water. Part I. Hydrolytic and hydrothermal processes. *J. Supercrit. Fluids* **2009**, *47*, 373–381. [CrossRef]
47. Hodge, J.E. Dehydrated Foods, Chemistry of Browning Reactions in Model Systems. *J. Agric. Food Chem.* **1953**, *1*, 928–943. [CrossRef]
48. Moreschi, S.R.M.; Petenate, A.J.; Meireles, M.A.A. Hydrolysis of ginger bagasse starch in subcritical water and carbon dioxide. *J. Agric. Food Chem.* **2004**, *52*, 1753–1758. [CrossRef] [PubMed]
49. Sjöström, E. *Wood Chemistry. Fundamentals and Applications*; Academic Press, Inc.: San Diego, CA, USA; London, UK, 1993.
50. Yang, H.; Yan, R.; Chen, H.; Lee, D.H.; Zheng, C. Characteristics of hemicellulose, cellulose and lignin pyrolysis. *Fuel* **2007**, *86*, 1781–1788. [CrossRef]
51. Kruse, A.; Zevaco, T. Properties of Hydrochar as Function of Feedstock, Reaction Conditions and Post-Treatment. *Energies* **2018**, *11*, 674. [CrossRef]
52. Liu, C.; Huang, X.; Kong, L. Efficient Low Temperature Hydrothermal Carbonization of Chinese Reed for Biochar with High Energy Density. *Energies* **2017**, *10*, 2094. [CrossRef]
53. Ekpo, U.; Ross, A.B.; Camargo-Valero, M.A.; Williams, P.T. A comparison of product yields and inorganic content in process streams following thermal hydrolysis and hydrothermal processing of microalgae, manure and digestate. *Bioresour. Technol.* **2016**, *200*, 951–960. [CrossRef] [PubMed]
54. Lucian, M.; Volpe, M.; Gao, L.; Piro, G.; Goldfarb, J.L.; Fiori, L. Impact of hydrothermal carbonization conditions on the formation of hydrochars and secondary chars from the organic fraction of municipal solid waste. *Fuel* **2018**, *233*, 257–268. [CrossRef]
55. Hoekman, S.K.; Broch, A.; Robbins, C. Hydrothermal Carbonization (HTC) of Lignocellulosic Biomass. *Energy Fuels* **2011**, *25*, 1802–1810. [CrossRef]
56. Bicker, M.; Endres, S.; Ott, L.; Vogel, H. Catalytical conversion of carbohydrates in subcritical water: A new chemical process for lactic acid production. *J. Mol. Catal. A Chem.* **2005**, *239*, 151–157. [CrossRef]
57. Rasrendra, C.B.; Makertihartha, I.G.B.N.; Adisasmito, S.; Heeres, H.J. Green Chemicals from D-Glucose: Systematic Studies on Catalytic Effects of Inorganic Salts on the Chemo-Selectivity and Yield in Aqueous Solutions. *Top. Catal.* **2010**, *53*, 1241–1247. [CrossRef]
58. Mäki-Arvela, P.; Simakova, I.L.; Salmi, T.; Murzin, D.Y. Production of lactic acid/lactates from biomass and their catalytic transformations to commodities. *Chem. Rev.* **2014**, *114*, 1909–1971. [CrossRef] [PubMed]
59. Zan, Y.; Sun, Y.; Kong, L.; Miao, G.; Bao, L.; Wang, H.; Li, S.; Sun, Y. Formic Acid-Induced Controlled-Release Hydrolysis of Microalgae (Scenedesmus) to Lactic Acid over Sn-Beta Catalyst. *ChemSusChem* **2018**. [CrossRef] [PubMed]
60. Aida, T.M.; Ikarashi, A.; Saito, Y.; Watanabe, M.; Smith, R.L.; Arai, K. Dehydration of lactic acid to acrylic acid in high temperature water at high pressures. *J. Supercrit. Fluids* **2009**, *50*, 257–264. [CrossRef]
61. Steinbach, D.; Kruse, A.; Sauer, J.; Vetter, P. Sucrose Is a Promising Feedstock for the Synthesis of the Platform Chemical Hydroxymethylfurfural. *Energies* **2018**, *11*, 645. [CrossRef]
62. Licursi, D.; Antonetti, C.; Fulignati, S.; Vitolo, S.; Puccini, M.; Ribechini, E.; Bernazzani, L.; Raspolli Galletti, A.M. In-depth characterization of valuable char obtained from hydrothermal conversion of hazelnut shells to levulinic acid. *Bioresour. Technol.* **2017**, *244*, 880–888. [CrossRef] [PubMed]

Article

Key Development Factors of Hydrothermal Processes in Germany by 2030: A Fuzzy Logic Analysis

Daniel Reißmann [1,*], Daniela Thrän [1,2] and Alberto Bezama [1]

[1] Helmholtz-Centre for Environmental Research – UFZ, Permoserstraße 15, 04318 Leipzig, Germany; daniela.thraen@ufz.de (D.T.); alberto.bezama@ufz.de (A.B.)

[2] Deutsches Biomasseforschungszentrum gemeinnützige GmbH, Torgauer Straße 116, 04347 Leipzig, Germany

* Correspondence: daniel.reissmann@ufz.de (D.R.); Tel.: +49-341-235-1267

Received: 9 November 2018; Accepted: 14 December 2018; Published: 19 December 2018

Abstract: To increase resource efficiency, it is necessary to use biogenic residues in the most efficient and value-enhancing manner. For high water-containing biomass, hydrothermal processes (HTP) are particularly promising as they require wet conditions for optimal processing anyway. In Germany, however, HTP have not yet reached the industrial level, although suitable substrates are available and technological progress has been made in previous years. This study aims to determine why this is by identifying key factors that need to occur HTP development in Germany until 2030. By using results of previous analyses within this context (i.e., literature review, SWOT analysis, expert survey, and focus group workshop) and combining them with the results of an expert workshop and Delphi-survey executed during this analysis, a comprehensive information basis on important development factors is created. Fuzzy logic is used to analyze these factors in terms of interconnections, relevance, and probability of occurrence by 2030. The results show that technological factors, such as a cost-efficient process water treatment and increased system integration of HTP into bio-waste and wastewater treatment plants, are given high relevance and probability of occurrence. The adaptation of the legal framework, for example, the approval of end products from HTP as standard fuels, has very high relevance but such adaptions are considered relatively unlikely.

Keywords: hydrothermal processes; Germany; fuzzy Delphi method; fuzzy logic cognitive map

1. Introduction

The German government has set a target of reducing the country's annual greenhouse gas emissions (GHG) by 50% in 2030 compared to the 1990 level [1]. To achieve this goal, it is necessary to use scarce resources more sustainably, which also includes a more efficient use of biogenic residues. However, currently, considerable amounts of biogenic residues and waste are being inefficiently used or not used in Europe [2,3]. The treatment of wet and sludgy biomass is particularly challenging, as it requires energy- and cost-intensive pre-treatment processes (e.g., drying, thickening, sanitization) to become suitable for conventional biomass treatment paths (e.g., pyrolysis) [4]. However, to enhance resource efficiency by sustainably utilizing residues and therefore, fostering progress towards a circular and bio-based economy, it is worth striving for value-added use of such materials. This could also reduce costs (e.g., for more expensive primary materials) and GHG (e.g., by substituting the energetic use of fossil resources), save scarce natural resources (e.g., by recycling of nutrients like phosphorus out of the residual flows) and thus promote climate protection [5–8].

For the last few years, hydrothermal processes (HTP) have gained attention as promising technologies to manage wet biomass. HTP transform wet substrates into gaseous, liquid, or solid high carbon and energy containing products via thermochemical conversion. The products can be used for several purposes, like direct use for energy production or as an intermediate for producing agricultural and pharmaceutical chemicals [4,9,10]. For optimal operation, HTP need high water containing substrates, which is why residues like sewage sludge and animal excreta are particularly suitable [9,11].

Depending on the operational conditions, different HTP types occur. At temperatures between 160 and 250 °C, pressure conditions between 10 to 30 bar, and a residence time between 1 to 72 h, hydrothermal carbonization (HTC) takes place. HTC is a coalification process that converts biomass into hydro-char [12] to be used for energetic purposes, material applications, and as fertilizer or soil conditioner [13]. At slightly higher temperatures (180 to 400 °C) and pressures (40 to 200 bar) but lower residence times (10 to 240 min), hydrothermal liquefaction (HTL) occurs. HTL is a process that transforms biomass into chemicals and bio-oils [14]. The products can be used for energy production and chemical industry [9]. At supercritical conditions (375 to 500 °C, 230 to 400 bar) hydrothermal gasification (HTG) takes place which usually needs less than 10 min for the reaction. Through HTG biomass is converted into gaseous materials, especially methane and hydrogen, which are used for energy and chemical industry [15].

Compared with other generally suitable biomass conversion processes (e.g., torrefaction, pyrolysis, composting), HTP have some advantages. Compared to torrefaction, for example, HTC products can achieve a higher energy density, energy yield, and combustion reactivity [16]. Additionally, HTC can provide economic advantages. For example, a comparative study of HTC, anaerobic digestion, and composting on the conversion of food waste showed that HTC performs economically best due to its low residence time and less substrate pre-treatment [17]. Another study showed that the HTL of algae can be advantageous compared to pyrolysis in terms of conversion yields and energy conversion rates [18].

At a first glance, HTP seem well suited to the conversion of wet biomass into high carbon and energy-containing products. Nevertheless, as a trade registry evaluation on HTP companies in Germany showed, so far, the technology has not prevailed in Germany. Based on this, since 2008, only a handful of new company foundations have been registered. This is in contrast with the general interest in these processes, which can be measured in terms of the level of research and technological progress. For example, according to a recent study, there are currently 15 patents on HTC in Germany [19]. Also, scientific interest in HTP is continuously increasing. According to Kruse and Dahmen [20], numerous published studies in Scopus since 2009 contain the keywords "supercritical gasification", "hydrothermal liquefaction", and "hydrothermal carbonization". This ongoing interest indicates that there is still high potential for HTP to become an innovative biomass conversion path. This has also been confirmed by international developments. Research activities on HTP are a core issue of the Pacific Northwest National Laboratories in the U.S., where some pilot plants are also in operation [21–23]. In addition, TerraNova Energy operates a larger HTC plant in China [24] and Ingelia in Spain [25].

Also, key metrics on HTP (e.g., the higher heating value (HHV) of products, the energy and mass balance of processes, the carbon efficiency, and the specific investment and operating costs) indicate that there is potential for HTP to be further developed at a large scale. For example, the HHV of hydro-coal ranges from 24 MJ/kg (median) to 26 MJ/kg (maxima) [25–27]. In terms of the energy efficiency of HTC (including all energetic losses during the process and without a utilization step) there is also high variation—between 62 per cent (median) and 77 per cent (maximum) [28–30].

However, optimization of the technological, economic, and ecological features of HTP depends on many parameters, such as heat recovery, applied catalysts, substrates used and their moisture content, logistics as well as plant sizes [4]. An example is the connection between HTC plant sizes and investment costs based on the manufacturers' information. The specific investment costs tend to decrease in relation to the capacity of the plants per additional ton of fresh matter biomass input (from 260 EUR/ton for 5000 tons capacity up to 50 EUR/ton for 80,000 tons capacity) [31–34]. So, economies of scale can be already observed. Further, learning curve and scale effects through more experience in the operation of plants on an industrial scale are crucial to achieving gradual optimization of essential parameters. Finally, if the parameters can be optimized, HTP will provide several advantages.

For example, the HHVs of the final products are generally higher than those of fossil reference systems [4]. Greenhouse gas savings compared to fossil references may also be significant, depending on the substrate used, the energy balance, and the subsequent product use [4].

So far, only a few studies have provided information on the future development of HTP in Germany and Europe as well as the corresponding key factors. A study of the German National Academy of Science and Engineering analyzed the potential system contributions of HTC and HTL to the flexibility of a renewable energy system until 2023 in Germany [35]. It was identified that the approval of HTC coal as a standard fuel and a corresponding fuel standard are of high importance. Furthermore, they recommended the promotion of nutrient recycling and the development of a cost-effective process water treatment procedure. They suggested the use of hydro-coal as an energy carrier, soil additive, and industrial carbon carrier. For HTL it is considered critical that in Germany, algae, which is a particularly suitable substrate, is largely missing. Nevertheless, they recommend the support of nutrient recycling and the increase in quality of the liquid product [35]. De Mena Pardo et al. [19] outlined the necessary factors for the establishment of HTC at the European level, such as the abolition of the waste status of HTC products from waste biomass. They predicted that hydro-coal will first become established on the energy markets and, in the long term, will also occupy material markets. In terms of establishment in the energy sector, however, the "end of the waste" characterization is crucial. Another recent paper [20] identified the integration of HTP into bio-refineries as important future development strategy to generate synergies. Furthermore, the whole value-chain must be addressed, also including stakeholders who have so far only been marginally involved, like farmers. In a previous paper, we used a SWOT analysis to identify the most important current barriers and possibilities for HTP in Germany [36]. The results indicated that the technological readiness of the plant, including the presence of high energy and material efficiency as well as the presence of a suitable process water treatment procedure are factors of high importance. In addition, the overall costs for producing the end-product and the competitive nature of sales markets are seen as important threats. Also, the GHG are of high relevance throughout the process and can be primarily viewed as an opportunity if HTP can mobilize their potential for emission savings as compared with fossil reference systems.

However, although HTP has some promising features as a resource efficient conversion technology for wet biomass, no scaling-up is happening in Germany. Thus, this study aims to identify and prioritize key development factors for HTP that should occur in Germany by 2030 and points out their interconnections using a structured expert participation process. Furthermore, the probability of occurrence of these factors is estimated. This study also aims to provide important information on barriers that must be dealt with to allow HTP to contribute to climate and resource protection in the future.

Specifically, we used the Fuzzy Delphi Method (FDM) and Fuzzy Cognitive Mapping (FCM) in this study. The Delphi method is a forecasting procedure based on the opinions of anonymous experts collected through a multi-stage survey process. It aims to systematically foster expert consensus about uncertain developments [37]. A Delphi survey consists of several rounds of interviews. The first round usually asks for the assessment of uncertain factors and events. The following rounds then ask the experts to revise or confirm their assessments based on the results of the previous rounds [38]. As this method contains some disadvantages (e.g., relatively low consistency of expert opinions, high enforcing effort, and sometimes modifications to individual opinions in order to reach consistent total opinions), we expanded it by using the Fuzzy Delphi Method (FDM) for the final evaluation. With FDM, expert opinions are integrated with fuzzy numbers based on the cumulative frequency distribution and fuzzy integrals. Thus, FDM applies triangulation statistics to determine the distance between the levels of consensus within the expert panel [39]. Furthermore, the FDM needs just a small survey panel to deliver reliable results—an advantage for studies with a small number of suitable participants [40]. FCM is a model consisting of nodes that indicate the most relevant factors (in FCM the term "concepts" is used) of a decisional environment and relationships between them (arcs and edges). The analytical background of FCM is based on the structure and function of concept maps, including graph theory-based analyses of pairwise structural relationships between the model factors. It is therefore a decision-support tool which originated a combination of fuzzy logic and artificial neural network theory [41]. It aims to define the important factors relevant to a specific community and the relationships between them as well as optionally testing scenarios in which these factors are varied to see how the system might react under a set of possible conditions [42]. An adjacency matrix A represents the interconnections between model factors. On that basis, the number and directions of edge relations are transformed into quantitative values between −1 (inhibitory effect) and +1 (positive effect) [43]. In particular, FCM can be used to model complex systems with high uncertainty and less available empirical data [44], which, based on our experiences within this working field, is the case for this study's topic.

2. Materials and Methods

The key factors were primarily developed based on qualitative and quantitative expert evaluations and information from relevant literature. Figure 1 gives an overview of the study design.

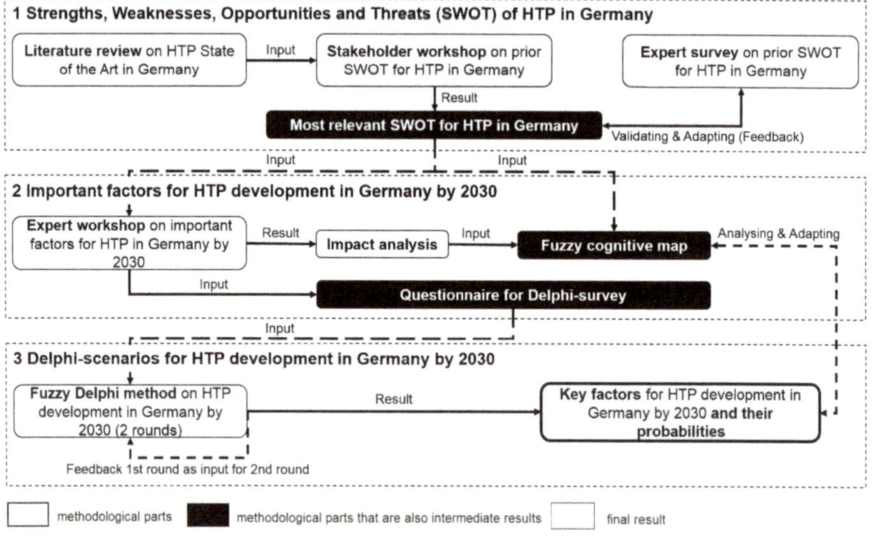

Figure 1. Study design.

The methodological framework is, in part, similar to the Hybrid Delphi method [45]. However, it also includes further methodological elements (literature review, impact analysis, fuzzy logic). Based on a comprehensive literature study [4], a moderated focus group workshop on the success and risk factors of HTP development in Germany was carried out. The results were validated and underpinned by a subsequent expert survey. A total of 41 experts, primarily scientists, plant manufacturers and plant operators from Germany and Switzerland, participated in the workshop. The expert survey panel consisted of feedstock suppliers, technology developers, technology users, retailers, product users, policy makers, and researchers from Germany. Within the workshop, the experts were asked about certain success and risk factors for HTP in Germany that were then collected, categorized, and discussed. In a subsequent expert survey, the results of the workshop were further validated by asking about the strengths, weaknesses, opportunities, and threats for HTP development in Germany. For the detailed procedure and the results of the SWOT analysis, see [36].

Based on these initial findings, a "long list" of important factors of HTP future development, their relationships, and interactions was derived through an expert scenario workshop. Six HTP researchers from the German Biomass Research Centre (DBFZ) participated. The influence analysis performed in this step served as the basis for the development of a Fuzzy-logic Cognitive Map (FCM), which provides an overview of all identified factors/concepts and their relationships. To construct the FCM, however, further expert feedback from the surveys and information from the literature review were included. In this analysis, we used multiple-valued logic scalar numbers from the discrete set $\{-1; -0.5; 0; +0.5; +1\}$ to determine the impact relations (arcs and edges) between FCM nodes (concepts). The open source web-based application Mental Modeler was used to create the FCM and identify the factors/concepts importance and connectedness [46].

Based on the results of the expert workshop and the FCM, a questionnaire for a Delphi survey was compiled and sent to 51 HTP experts via an online survey. The FCM factors/concepts (Appendix A, Table A1) served as essential inputs for the preparation of the Delphi questionnaire. However, the use of too many survey items makes cognitive assessments more difficult and thus tends to reduce the reliability of the results, which is why it was decided to integrate particularly factors/concepts with a high FCM centrality (cf. Table 1) into the survey. Nevertheless, following feedback received during the first round of interviews, several items were added to the second questionnaire.

The survey participants were selected based on their expertise. Selection criteria were as follows: (1) academic or professional recommendations, (2) well-known authors of relevant publications on the specific subject, (3) stakeholder group representative, and (4) estimated professional experience within the working field. These criteria were selected based on the suggestion by Stevenson [47] and Hasson et al. [48] to mainly include experts in the field of study (indicated through criteria 1, 2 and 4) as well as different stakeholders (criterion 3). The international participants were asked about developments of HTP in the European context, since they were assumed to have, at best, limited knowledge on the German situation. However, both the German and the European situations are comparable. Figure 2 gives an overview of the composition of the participants, their expertise, and the nations represented in the first round of interviews. The relative distribution in the second round of the survey (n = 12) was very similar.

Two rounds were conducted in this study. Twenty-seven experts participated in the first round (response rate 1st round: 53%). Of these 27 people, twelve participated in the second round (response rate 2nd round: 44%). The following item-categories were part of the survey (assessment scales are explained in the Appendix B, Table A2): (1) relevance of factors for HTP development in Germany by 2030, (2) relevance of risks for HTP development in Germany by 2030, (3) estimated probabilities of factor occurrence by 2030 and (4) certainty in assessing per item-category.

Besides evaluating with scales, the experts had the opportunity to explain their selection and assessment in text fields. For both rounds, 22 comments on capacity development, five comments on success factors, four notices regarding risks, and eleven notes on the development of biomass utilization rates were provided. By means of qualitative content analysis (i.e., differentiation between pros and cons, frequencies of keywords, identification of consensus statements) essential statements were summarized (Appendix C, Table A3). The hints of the first round were also included in the preparation of the questionnaire for the second round.

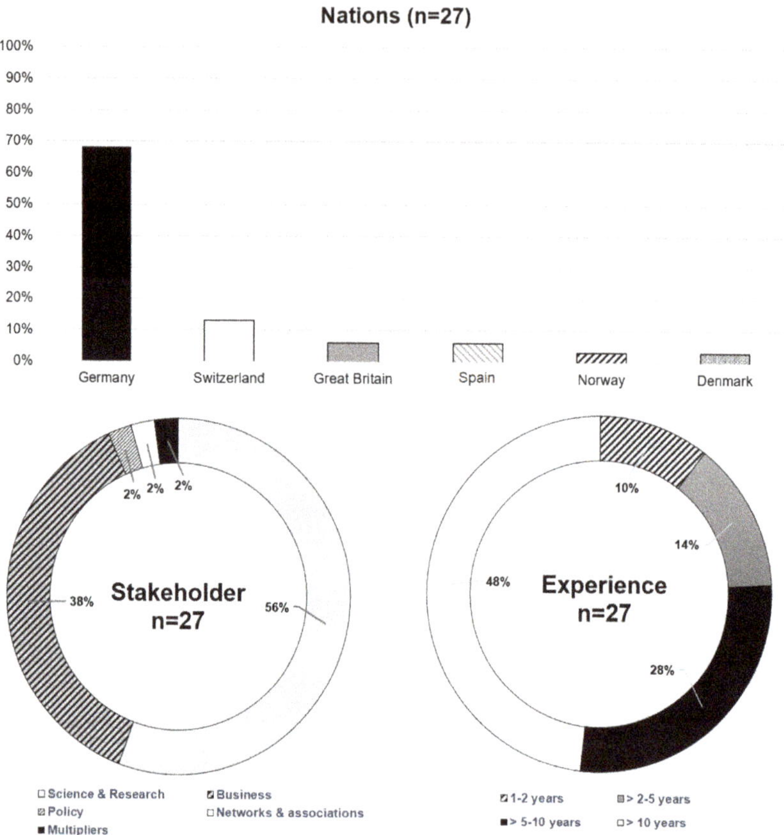

Figure 2. Participants of the first Delphi-survey round.

After the first round, an interim evaluation took place, which showed the degree of agreement in the expert assessments and the frequency of distributions of the first tendencies by descriptive statistics (median, standard deviation, interquartile range (IQR)). The questionnaire for the second round of the survey was adjusted, taking into account the results from round 1. After executing the Delphi survey, we analyzed the results by using the FDM which consists of the following steps [49]:

1. Determining experts (see previous explanations).
2. Selecting a linguistic scale to be converted into a fuzzy-scale (cf. Appendix B, Table A2).
3. Calculating the difference between the average fuzzy number (m) and each experts' fuzzy number (n) per item by using following formula:

$$d\,(\widetilde{m},\,\widetilde{n}) = \sqrt{\frac{1}{3}\left[(m1-n1)^2+(m2-n2)^2+(m3-n3)^2\right]} \quad (1)$$

4. Determining the threshold value for consensus/dissent of the expert panel:

In accordance with [38], we chose a threshold of $d \leq 0.2$ to make a decision as to whether the experts had reached consensus on the item. Next to this, the frequency of expert agreement is presented as the percentage of $d \leq 0.2$ per item-category in relation to all items. A value of $\leq$75% represents panel consensus.

5. Defuzzification:

To determine a ranking of the most relevant/probable factors per item-category, it is necessary to defuzzify the fuzzy values into a crisp-value (A_i). For this, we used the following formula in accordance with [38]:

$$A_i = \frac{1}{3}(m1 + m2 + m3) \quad (2)$$

3. Results

3.1. Factors for HTP Development in Germany by 2030 and Their Relations

The development factors and risks were primarily derived on the basis of the expert workshop and the aforementioned previous SWOT analysis executed by the authors [36]. Above all, the expert workshop served as the basis for identifying areas of interest. The factors were then further differentiated and backed up with information from the literature. The whole list of factors is part of the appendix (Table A1).

The factors were assessed in the expert scenario workshop by means of an impact matrix with regard to their mutual influences. Based on this, a Fuzzy-logic Cognitive Map (FCM) was constructed. Figure 3 shows a part of the overall FCM for the relationships of the factor "Regular Fuel Recognition".

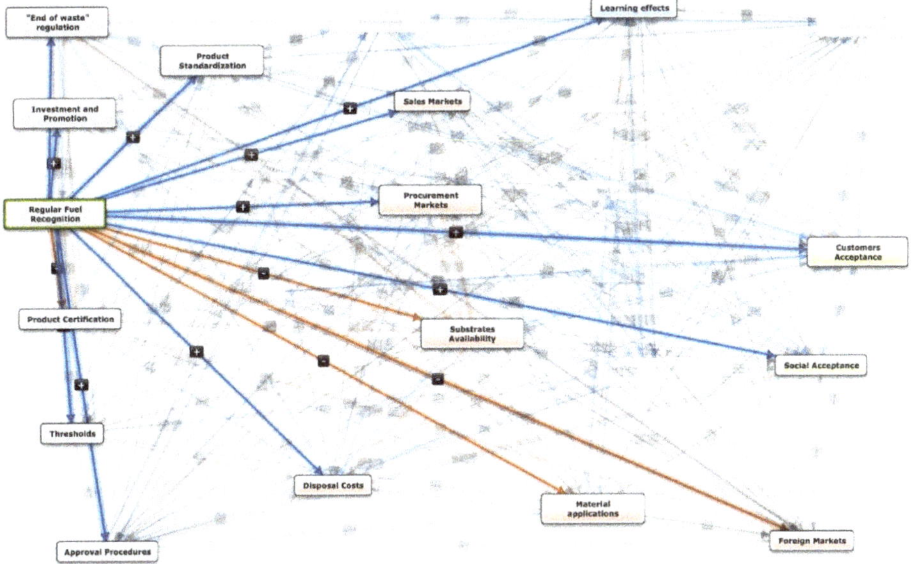

Figure 3. Part of the Fuzzy-logic Cognitive Map (FCM) for the impact of the concept "Regular Fuel Recognition" on other system concepts (expert knowledge-based FCM created with the Mental Modeler).

Since FCM is based on graph theory, which provides a wide variety of indices, we can also make statements about the structure of the system as well as gain information about the functions of individual factors. Table 1 lists the most relevant metrics for the developed FCM including a short definition of the indices.

Table 1. FCM indices and their scores in the hydrothermal processes (HTP) model.

FCM Indices	Explanation	Indices in HTP Model
N (concepts)	Indicates the total number of system factors [50,51].	24
N (connections)	Indicates the total number of connections between the system concepts [50,51].	235
N (transmitters)	Indicates the total number of concepts that influence other concepts but are not affected by other concepts [50,52].	0
N (receiver)	Indicates the total number of concepts that are influenced by other concepts but have no effect on them [50,52].	1
N (ordinary)	Indicates the total number of concepts that affect and are affected by other concepts [50,52].	23
Density	This index shows the networking degree of the system, i.e., the number of concepts and edge relations. A high density indicates that several probable management options exist [50,52]. The density can have a value between 0 and 1.	0.43
C/N	The number of connections divided by the number of concepts. A low C/N score indicates a high degree of system connectedness [50,51]. Low is relative in this context, because it must be seen in context with other comparable systems.	9.79
Outdegree & Indegree	Information about the concept degree as a transmitter (driver), receiver (output), or force that conveys effects (ordinary) [53].	Highest Outdegree (highest driving force): Regular Fuel Recognition (11.5)
		Highest Indegree (highest receiving force): Customer Acceptance (10)
Centrality	Indicates how strongly a concept influences the whole system [53,54].	Highest Centrality: Regular Fuel Recognition (13.5)
Complexity	Illustrates the degree of model accuracy and measures the degree to which outcomes of driving forces are considered [50,51].	Infinite

3.2. Results of the Fuzzy-Delphi Method

Table 2 summarizes the results of consensus or dissent after the second round. For this purpose, the determined fuzzy values (d) are given, where $d \leq 0.2$ is the threshold value. Grey shaded values are the factors where consensus was reached. In addition, the percentage of expert consensus is specified. This indicates how many item evaluations of the entire panel in relation to the total items did not exceed the threshold. Here, a value of at least 75% is the consensus criterion.

Table 2. Results on the fuzzy evaluation regarding expert consensus/dissent after round 2.

No.	Thematic Category	Consensus/Dissent after Round 2 (n = 12)		
		d_{factor}	d_{risk}	$d_{probability}$
Political-legal factors				
1	Regular fuel recognition	0.178	0.183	0.204
2	Investment and promotion	0.277	0.241	0.136
3	"End of waste" regulation	0.170	0.221	0.263
4	Product certification	0.153	0.221	0.164
5	Thresholds	0.300	0.239	0.288
6	Approval procedures	0.267	0.239	0.236
7	Product standardization	0.204	0.170	0.136
8	Substrate standardization	0.159	-	0.192
9	Process standardization	0.083	0.265	0.213
Economic factors				
10	Sales markets	0.187	0.085	0.181
11	Procurement markets	0.209	0.170	-
12	Substrate availability	0.187	0.186	0.166
13	Disposal costs	0.209	0.293	0.199
14	Material applications	0.226	-	0.235
15	Foreign markets	-	0.208	-
Technological factors				
16	Process water treatment	0.170	0.204	0.136
17	System integration 1	0.115	-	0.162
18	System integration 2	0.229	-	0.187
19	Nutrient recycling	0.178	-	0.236
20	Learning effects	0.200	0.140	0.200
21	Accidents	-	0.265	-
Ecological factor				
22	Life cycle performance	0.378	-	-
Mean d_i		**0.207**	**0.205**	**0.193**
Percentage of expert consensus		72%	71%	76%

Table 2 shows that after the second survey round, majority consensus was achieved in at least one item-category (factors, risks, probabilities). However, for thresholds, approval procedures, material applications, foreign markets, accidents, and life cycle performance, no consensus was reached at all. The panel consensus (last row of Table 2) was not reached regarding factors and risks (<75%), which likely shows that the expert assessments tended to be furthest apart for these item-categories.

However, compared to the first round, the second round showed a significant increase in expert consensus. The expert consensus rate increased by 28 percentage points in the assessment of the relevance of the factors, by 19 percentage points in the assessment of the relevance of the risks, and even, by 33 percentage points in the probability of occurrence estimates between the rounds. For some items, there were considerable differences. In particular, the relevance of process standards showed a very strong difference between rounds 1 and 2 ($\Delta d_{factor} = -80\%$). This could be due to the fact that in the second round, experts who regard process standards as equally relevant in particular were still involved. This reveals one of the weaknesses of the Delphi method, as there is sometimes a high drop-out rate (in this case 56%) between the rounds that can cause changes in the results due to differences in the survey panel, rather than solely due to adjustments based on the previous round's results. However, one basic assumption of the Delphi method is that expert consensus increases due to the adaption of evaluation based on the previous round's results, which is why we basically also assumed this for the consensus increase in this study. For the factors/concepts in which a consensus was reached (grey shaded in Table 2), Table 3 shows the values (A_i) after defuzzification. Based on this, the items' fuzzy logic-based relevance/probability can be ranked. Factors/concepts that are not greyed out in Table 3 were no longer considered in the corresponding categories, as a dissent prevailed in the expert assessments. We differentiated between:

- A_f = defuzzified value for factors
- A_r = defuzzified value for risks
- A_p = defuzzified value for probabilities
- A_c = defuzzified value for certainty in assessment
- $Rank_f$ = Rank in relation to other factors
- $Rank_r$ = Rank in relation to other risks
- $Rank_p$ = Rank in relation to other probabilities

Table 3. Ranking of consensus items in terms of relevance and probabilities after defuzzification.

No.	Factors with Consensus in at Least One Item-Category	A_f	$Rank_f$	A_r	$Rank_r$	A_p	$Rank_p$
Political-legal factors							
1	Regular fuel recognition	8.2	3	5.8	2	n.c.	n.c.
2	Investment and promotion	n.c.	n.c.	n.c.	n.c.	2.9	9
3	"End of waste" regulation	8.6	2	n.c.	n.c.	n.c.	n.c.
4	Product certification	7.4	5	n.c.	n.c.	3.8	8
7	Product standardization	n.c.	n.c.	5.6	3	3.9	7
8	Substrate standardization	2.6	10	-	-	2.0	10
9	Process standardization	2.8	9	n.c.	n.c.	n.c.	n.c.
Economic factors							
10	Sales markets	4.6	8	2.8	5	6.1	4
11	Procurement markets	n.c.	n.c.	4.4	4	-	-
12	Substrate availability	5.0	7	2.8	5	6.8	1
13	Disposal costs	n.c.	n.c.	n.c.	n.c.	6.2	3
Technological factors							
16	Process water treatment	8.0	4	n.c.	n.c.	6.8	1
17	System integration 1	9.0	1	-	-	6.0	5
18	System integration 2	n.c.	n.c.	-	-	4.6	6
19	Nutrient recycling	8.2	3	-	-	n.c.	n.c.
20	Learning effects	6.4	6	7.4	1	6.4	2
Certainty in the assessment of the item category according to the experts' own statements: A_c		6.4		5.6		5.0	

"n.c." = no consensus reached; "-" = factor was not part of this item-category.

Table 3 shows that the assessments of the relevance of occurrence of a factor (A_f) and the risk of non-occurrence (A_r) are very different. For example, the absence of learning effects (e.g., lack of reference facilities) is considered to be a significant risk (7.4). However, the relevance of this factor is also still high (6.4) but only in the midfield relative to other factors. The uncertainty according to the panelists' own assessments (A_c) is highest in the probabilities and lowest in the relevance of the factors. However, the values are close to each other, which is why the assessment certainty of the item categories is largely the same.

Regarding the relationships between mutually relevant factors and corresponding probabilities (grey shaded in Table 3), only a few factors show high values (i.e., near to 10) for both. Figure 4 visualizes the relationships.

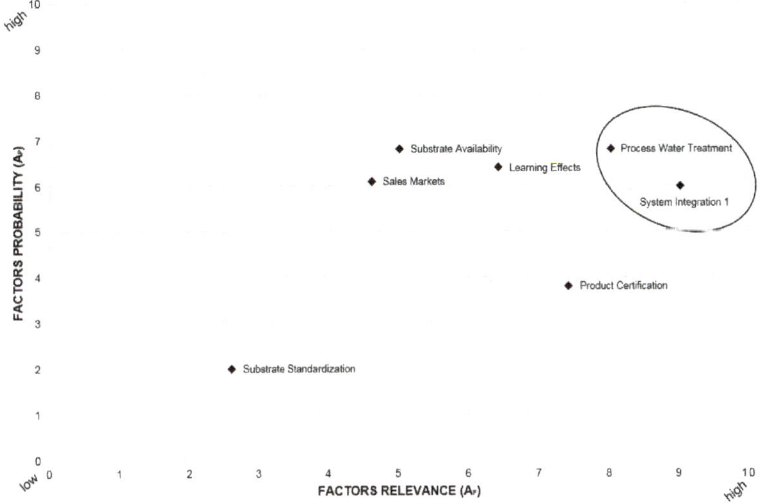

Figure 4. Combination of relevance and probability of consensus factors after defuzzification.

Only two factors were considered to be highly relevant and also highly probable. Namely, the introduction of a cost-effective process water treatment and the integration of HTP into existing bio-waste and wastewater treatment plants.

4. Discussion

Although HTP has already been shown to have an advantage on some points (e.g., HHV, energy yields, decreasing specific investment costs while increasing capacity), the analysis showed that there are several factors related to the development of HTP in Germany that have hindered successful development so far. Above all, political-legal aspects are strongly inhibiting a scale-up in Germany, but adaptions in the near future are considered unlikely. This shows that the experts involved think that the legislator or the political decision-makers have relatively little ambition to promote the development of HTP more strongly. This is already evident today as some German HTP plant manufacturers and operators are already focusing on foreign markets (especially China). Nevertheless, HTP could considerably contribute to the achievement of a bio-based economy by efficiently converting currently difficult-to-use wet biomasses into valuable products. However, the adaptation of the legal framework is urgently needed for this. If the national legislator does not take action, an important step could also be the development of an EU regulation on the end-of-waste status of waste biomass products, similar to those already introduced for scrap iron, scrap steel, and scrap aluminum as well as for certain types of glass. One of the reasons for this is that the legal uncertainty for plant operators and product users is very high, which, in turn, increases transaction costs [55]. Due to the fact that HTP products cannot be used as standard fuels, the energy market cannot be fully penetrated, which significantly reduces the product's market potential. However, there are still many problems at the technological level. So far, Germany is still a technology leader in the field of HTP (e.g., as indicated through patents) [19]. Based on results of this analysis, it is politically recommendable to work actively on measures that ensure that HTP are used economically in Germany and do not become exclusively an export product as this could cause related companies to relocate their headquarters abroad.

In addition, technological advancements are considered to be relevant drivers and are also estimated to be relatively likely. Above all, a mature technology for the cost-effective treatment of the process water is urgently needed to reduce the overall related costs and thus increase the cost-effectiveness of the process. In addition, an efficient treatment process for polluted water is also needed to aid in environmental protection. Potential for promoting the development of HTP is seen particularly in system integration, for example, into existing bio-waste and waste-water treatment plants (WWTP). The resulting synergies can, in particular, save logistics costs and directly link the locations of substrate occurrence, conversion technology and, in some cases, customers. The experts probably regard technological advances as likely because corresponding research and development is very active. In particular, cost-effective solutions for the process water treatment are being intensively researched [56–58], which is why suitable solutions are likely to be expected in this area in the foreseeable future. As an overview, Table 4 summarizes the main results of this study, i.e., the most crucial barriers and potential for future HTP development and the spread of technology as well as suggestions for possible measures to reach the potential benefits for HTP and reduce the barriers to achieving these.

Table 4. Consensual key potential benefits and barriers for HTP development in Germany by 2030 and potential measures.

Key Development Factor(s)	Potential Measure(s) to Reach Potential Benefits or Reduce Barriers
Key potentials	
Political-legal	
An end-of-waste regulation is being introduced for HTP products (i.e., products from bio-waste, sewage sludge etc.), and HTP energetic products (e.g., hydro-coal) are recognized as standard fuels.	The European or national legislation has to be adjusted accordingly. This means that a regulation must be introduced that allows the energetic use of products from waste biomass. Such a regulation could be very similar to regulations already being introduced for broken glass and steel scrap.
Technological	
Integration of HTP into existing bio-waste treatment plants and waste-water treatment plants (WWTP) including nutrient recycling	Research on suitable technological solutions for the most efficient integration of HTP into such plants must be fostered. Concepts from biorefinery research could possibly be used as a basis for good solutions. However, relevant stakeholders, especially plant operators, must be closely involved (e.g., with common workshops) to reduce reservations and develop good concepts together. An important issue for bio-waste plant operators and WWTP operators could be nutrient recycling as this would provide an additional economic product (next to HTP products itself), which is highly demanded (esp. phosphorus [59])
Key barriers	
Political-legal	
Unambitious politics and obstructive legislation, i.e., no introduction of "end of waste" directive or alternative (e.g., product certification).	Relevant political decision-makers have to be motivated for legislative action. Scientifically-based policy advise (e.g., Scientific Advisory Boards) could be an important instrument to motivate decision-makers. To create a suitable argumentative basis for this, research on the economic and ecological benefits of HTP is necessary but must also be translated into easily understandable messages and communicated most efficiently. Next to this, political decision-makers must be integrated into several activities on HTP to increase attention on the technology. Best-practice cases (business cases) could also be useful to show the functioning and advantages of the technology.
Technological	
The understanding and knowledge of the process will not increase considerably (missing learning effects, for example, through missing reference systems/business cases).	To reduce this barrier, investment and promotion activities are especially important (e.g., by public or private funders and investors). Through this, larger pilot and demonstration plants can also be developed which may help to increase the understanding of the processes on larger scales. Such reference plants are important to give investors an impression of how the technology works, which, in turn, could generate further investments. Learning effects will occur if sufficient experience with the operation of larger plants is made. Business cases can serve as important information basis for new projects.

As mentioned in the introduction, few studies have focused on this issue so far. However, the results of this study are in line with the findings of the similar ones (e.g., the importance of having an efficient process water treatment procedure and the approval of HTP products from residues and waste as standard fuels) which confirms the high importance of the identified key factors. The novel aspect of this study, however, is that in addition to the relevant literature, extensive expert knowledge was included and evaluated in a structured manner. In addition, this study initially depicted all relevant key factors and did not focus on selected aspects directly at the start of the analysis, which is why the methodology can be regarded as non-normative. The application of FCM shows, for the first time, how the individual factors are related. The use of fuzzy logic also takes into account the bias of qualitative assessments (e.g., due to different participants' estimations of "important" and "unimportant"). Although the studies mentioned in the introduction showed very similar results to this analysis, some only considered individual technologies and not the entire technology platform (e.g., [19]) or they focused on very specific contexts (e.g., the contribution of HTC and HTL to the flexibility of a renewable energy system) (e.g., [35]), which is why not all relevant system factors were considered. The aforementioned studies did not prioritize the potential benefits and barriers to HTP development like this analysis, but they also classified them into categories and highlighted the high importance of the already mentioned legal and technological factors. Hence, this study confirms the entirety of the results of the mentioned studies and substantiates them both in terms of content (expert knowledge) and by using an alternative methodology (fuzzy logic).

The applied methodology to derive particularly relevant factors, risks, and probabilities of occurrence is unique in this form. Although other technology assessments have applied the Fuzzy Delphi method [60], Fuzzy-logic Cognitive Mapping [61], or SWOT analysis [62], they did not use such a combination. The advantage of this method is the versatile participation format that greatly increases the objectivity of the results overall, since several correction and feedback loops are part of it. The combination of workshops and surveys within this study makes it possible for both conduction of the discourse (workshops) and collection of anonymized content (Delphi survey) to occur. Although other comparable studies also applied participation as a qualitative methodological element [63], the particular kind of methodological combination used (cf. Figure 1) has not previously been used in the literature. The core of information filtering into relevant and probable factors is the Fuzzy Delphi Method. With a total of 27 experts from different stakeholder groups in the first round, this Delphi survey achieved a high level of representativeness, since there are very few HTP experts in the study area anyway. The number of participants is an extremely important factor in achieving meaningful results, so it is strongly recommended that experts are already mobilized before a study of this type is begun. Through the use of fuzzy logic, it became possible to bypass some disadvantages of the classical Delphi method. In particular, the different types of assessment by people on the basis of linguistic scales can be easily circumvented by fuzzy scales [64]. Another key element of this analysis was the application of the FCM method. Again, fuzzy logic was used to translate qualitative expert assessments into a model that represented the overall system of factors. In this study, the mapping was conducted as part of a workshop with six experts. We preferred a smaller group to ensure discussion and to prevent over-standardization of the workshop. A standardization of the mapping, for example, via online formats or targeted queries, would certainly allow a larger number of participants. The creation of an FCM requires a high level of cognitive performance, but it helps to structure the complexity of a system to identify feedback loops or so-called "hidden patterns". Identification of the dependencies of the factors must be carried out carefully, as this is the central way for the system effect to be identified. Nevertheless, the results are meaningful as a "scoreboard" and do not guarantee objective accuracy, as this is not the aim of a qualitative analysis like this one anyway. Looking into the future always involves high uncertainty and particularly shows ranges and opportunities.

5. Conclusions

In this study, we asked for the reasons why HTP does not yet prevail on a large industrial scale in Germany. By means of a literature- and expert knowledge-based fuzzy logic analysis, we identified key factors and prioritized them. The study results show that political and legal adjustments to the relevant framework conditions as well as technological improvements are seen as very important for the positive future development of HTP in Germany. This especially includes the key potential benefits shown in Table 4. These factors are strongly connected to other system components which shows their high impact on the whole system. The results can serve as important information for

HTP stakeholders in Germany, especially political decision-makers, entrepreneurs, and researchers. However, the limitations of the study are that the findings are only valid for the German situation. Other nations require their own comparative studies. Additionally, the study was highly qualitative in nature due to the insufficient information and data situation in this field of research. Hence, some uncertainty remains which is, nevertheless, very common for analyses that deal with future developments. In the future, the identified factors and interconnections shall serve as a basis for upcoming scenario case studies focusing on the system and plant levels (also, in part, quantitively). In this way, we hope to gain even more insight into desirable technological, economic, ecological, and political-legal developments for HTP by 2030 in Germany.

Author Contributions: Conceptualization, D.R. and A.B.; Data curation, D.R.; Formal analysis, D.R.; Funding acquisition, D.T. and A.B.; Investigation, D.R.; Methodology, D.R.; Project administration, D.T. and A.B.; Software, D.R.; Supervision, D.T. and A.B.; Validation, D.R.; Visualization, D.R.; Writing—original draft, D.R.; Writing—review & editing, D.R., D.T. and A.B.

Funding: This work was supported by the Helmholtz Association under the Joint Initiative "Energy System 2050—A Contribution of the Research Field Energy".

Acknowledgments: We are grateful to all experts, who have supported us within the various participation formats. Special thanks go to Benjamin Wirth, who helped to organize the expert workshop for the derivation of the impact matrix. Additionally, we want to thank the anonymous reviewers for their valuable and helpful comments and suggestions.

Conflicts of Interest: The authors declare no conflict of interest. The funders had no role in the design of the study; in the collection, analyses, or interpretation of data; in the writing of the manuscript, or in the decision to publish the results.

Appendix A. List of Relevant System Factors for HTP Development in Germany

The factors are formulated in positive form and thus represent a desired event; the corresponding negative formulation represents a risk. However, the non-occurrence of a factor is not always considered as a risk. In addition, risks were identified that do not necessarily represent a development factor in their inverse effect (accordingly, they are not formulated in positive form). Such factors are marked with asterisks.

Table A1. "Long list" of factors for HTP development in Germany by 2030.

x_i	Tagging	Factors/Concepts Explanation
Political-legal factors/concepts		
1	Regular fuel recognition	HTP energetic products (e.g., hydro-coal) are recognized as standard fuels. This factor is strongly connected to the fourth factor as this represents an alternative requirement for the recognition of HTP products as standard fuels.
2	Investment and promotion	Investment incentives (e.g., policy support instruments) and/or technology and research funding programs for HTP are being introduced or, rather, promoted.
3	"End of waste" regulation	An end-of-waste regulation is being introduced for HTP products (i.e., products from bio-waste, compost, etc.). Comparable regulations already exist for broken glass and steel scrap.
4	Product certification	Official recognition certificates for HTP products are introduced and issued accordingly by the competent authorities. This helps to reduce uncertainty in practice in terms of the classification of HTP products as fuels.
5	Thresholds	Thresholds relevant to HTP (e.g., the Federal Immission Control Act) are relaxed as far as reasonably possible.
6	Approval procedures	Approval procedures for new HTP plants are accelerated which might save costs during the planning and construction phase.
7	Product standardization	The quality of HTP products is standardized (e.g., fuel standard). This helps to reduce uncertainties with HTP products and sales markets (e.g., for product users) and enhances transparency.
8	Substrate standardization *	The quality of HTP substrates is standardized (e.g., ISO standard). This helps to reduce uncertainties with HTP procurement markets (e.g., for substrate users) and enhances transparency.
9	Process standardization	Process standards are introduced (e.g., ISO standard). This helps to reduce uncertainties for plant constructers and operators and enhances transparency.

Table A1. *Cont.*

x_i	Tagging	Factors/Concepts Explanation
Economic factors/concepts		
10	Sales markets	The competition on HTP relevant sales and product markets (e.g., energy carriers, fertilizers, substitutes for chemical products) decreases. Thus, the relative market share for HTP firms might be increased.
11	Procurement markets	The competition in HTP relevant procurement markets (e.g., animal excreta, sewage sludge) decreases. Thus, more usable substrates for HTP might be available, possibly near to the plant location.
12	Substrate availability	The available and technically usable amount of substrates increases. Thus, in centralized concepts, plants might be able to handle higher capacities, or in decentralized concepts, more substrates will be available near to the plant location assuming that substrate availability increases equally in Germany.
13	Disposal costs	Disposal costs for HTP substrates per mass unit (e.g., ton) are increasing. Thus, revenue for the disposal of such substrates might also increase which would generate additional income for HTP plant operators.
14	Material applications *	HTP products are primarily used for material applications (e.g., as fertilizer, functional carbon). This could result if energy markets remain unprofitable due to legal barriers (missing recognition as regular fuels). Products for HTP might be primary applied in markets for bio-based products. However, this factor strongly depends on missing legal adjustments regarding fuel recognition according to experts' opinions.
15	Foreign markets **	HTP plant manufacturers and operators concentrate almost exclusively on foreign markets. This might be a result of missing market demand, an insufficient or rather braking legal framework, low relative market shares for HTP products in related markets or missing political incentives and willingness to promote HTP in Germany.
Technological factors/concepts		
16	Process water treatment	A cost-efficient and sustainable solution for process water treatment is being developed and applied nationwide. This might promote the overall economic (and ecological) performance of HTP as the process water treatment is currently also a relevant cost (economic) factor that might make HTP concepts uneconomic.
17	System integration 1 *	HTP plants are increasingly being integrated into bio-waste and wastewater treatment facilities. Thus, the locations of substrate occurrence and treatment facilities could be integrated optimally, leading to lower logistic costs. Also, other synergies might be generated, e.g., process water is treated directly by the wastewater treatment plant on site.
18	System integration 2 *	HTP are increasingly being integrated into bio-refineries. This could also generate considerable synergies (e.g., cascade usage networks).
19	Nutrient recycling *	The nutrient recovery is enhanced. Especially, nutrient recovery from the process water might be promising as the process water must be treated anyway. Due to political and legal frameworks (2017 amendment of sewage sludge ordinance) that especially require phosphorus recovery from sewage sludge, this might be a useful strategy.
20	Learning effects	The process understanding and knowledge increases (learning effects, for example, through reference systems/business cases). According to the learning curve effect theory, this will especially reduce the cost per unit of product which is why this is also, in part, an economic factor [65].
21	Accidents **	Accidents with existing facilities reduce trust in the safety of the technology. This might especially affect plant operators and society which is why this factor is strongly connected to social factors.
Ecological factor/concept		
22	Life cycle performance *	Research on climate and resource protection by HTP will be intensified. Results on this will also successively improve the life cycle performance due to new insights (e.g., the stability of HTC coal in the soil as CO_2 sink). This might especially promote social acceptance of the technology. However, the life cycle performance is strongly connected to several other factors (e.g., reduced pollutants in process water after treatment) which is why this factor is just one part of promoting the life cycle performance.
Social factors/concepts		
23	Customer acceptance	Customer acceptance of HTP increases. This might be the result of technological progress, legal adjustments that promote HTP, higher transparency regarding HTP product quality (e.g., end-product customers), substrate quality, and process performance (e.g., customers for facilities/plant operators).
24	Social acceptance	The social acceptance of HTP increases or rather, society regards HTP as a resource efficient technology for future biomass conversion.

* Factor is not considered as a risk if it not occurs; ** Solely represents a risk.

Appendix B. Scale Relations

Table A2. Linguistic variables of Delphi survey item-categories and corresponding Likert and fuzzy scales.

Linguistic Scale	Likert Scale	Fuzzy Scale		
For item categories "relevance of factors" and "relevance of risks"				
extremely relevant	5	0.6	0.8	1
very relevant	4	0.4	0.6	0.8
relevant	3	0.2	0.4	0.6
barely relevant	2	0	0.2	0.4
irrelevant	1	0	0	0.2
For item category "probability of factors"				
very high	5	0.6	0.8	1
high	4	0.4	0.6	0.8
middle	3	0.2	0.4	0.6
low	2	0	0.2	0.4
very low	1	0	0	0.2
For item category "assessment (un)certainty"				
very certain	5	0.6	0.8	1
certain	4	0.4	0.6	0.8
relative certain	3	0.2	0.4	0.6
uncertain	2	0	0.2	0.4
very uncertain	1	0	0	0.2

Appendix C. Expert Statements in the Delphi Survey

Table A3. Summarized comments and hint of experts in the Delphi survey.

Category	Key Statements of the Experts
Arguments for a plant capacity increase	• Capacity will increase for plants that currently only exist on a pilot scale. • Capacity expansion due to legal adjustments and additional economic opportunities (e.g., additional revenue from rising carbon allowances due to an end of waste regulation for bio-coal). • Easy scalability of the systems due to modular design. • Learning effects, experience, and technological advances (for example, process water treatment solutions). • Scale effects and scale advantages. • HTC plants must be based on wastewater treatment plants of the size 3–4, therefore requiring a capacity of 50,000 metric tons biomass input per year.
Arguments against a plant capacity increase	• For the most relevant fields of HTP application (mainly the disposal sector), the current capacity is sufficient. • For wet biomass, only relatively small amounts are meaningful for ecologic (CO_2) and economic (costs) transport, which limits the capacity. • The plants are used decentral, because substrate availability is crucial. That limits the capacity.
Notes on relevant success factors	• HTP must be evaluated holistically to show its potential benefits. • Regulatory and political measures need to be implemented.
Notes on relevant risk factors	• Today's expectations of the technology will be not fulfilled (especially economically and ecologically). • The environmental effects are misjudged. • The pressure of competition is increasing.

Table A3. *Cont.*

Category	Key Statements of the Experts
Arguments for an increase in the biomass utilization rate	• Environmental benefits compared to landfilling and anaerobic digestion promote HTP deployment, but it has to be backed by legislation and incentives. • Growing environmental awareness.
Arguments against an increase in the biomass utilization rate	• No significant technological advancements. • Municipal users do not engage in HTP. • The spatial distribution of substrates limits their efficient use.

References

1. German Federal Government. *Deutsche Nachhaltigkeitsstrategie, Neuauflage 2016*; German Federal Government: Berlin, Germany, 2016; pp. 35–40.
2. Searle, S.; Malins, C. National Case Studies on Potential Waste and Residue Availability for Cellulosic Biofuel Production in the EU. The International Council on Clean Transportation, 2015. Available online: http://www.theicct.org/sites/default/files/ICCT_EU-national-wastes-residues_Feb2015.pdf (accessed on 6 October 2018).
3. Brosowski, A.; Thrän, D.; Mantau, U.; Mahro, B.; Erdmann, G.; Adler, P.; Stinner, W.; Reinhold, G.; Hering, T.; Blanke, C. A review of biomass potential and current utilisation—Status Quo for 93 biogenic waste and residues in Germany. *Biomass Bioenergy* **2016**, *95*, 257–272. [CrossRef]
4. Reißmann, D.; Thrän, D.; Bezama, A. Hydrothermal Processes as treatment paths for biogenic residues in Germany: A review of the technology, sustainability and legal aspects. *J. Clean. Prod.* **2018**, *172*, 239–252. [CrossRef]
5. German Federal Ministry of Food and Agriculture. *National Policy Strategy on Bioeconomy. Renewable Resources and Biotechnological Processes as a Basis for Food, Industry and Energy. Division 531—Strategy and Coordination of the Directorate-General 'Biobased Business, Sustainable Agriculture and Forestry'*; German Federal Ministry of Food and Agriculture: Berlin, Germany, 2014.
6. European Commission. *Innovating for Sustainable Growth: A Bioeconomy for Europe*; Communication COM(2012) 60 Final; European Commission: Brussels, Belgium, 2012.
7. Bezama, A. Let us discuss how cascading can help implement the circular economy and the bio-economy strategies. *Waste Manag. Res.* **2016**, *34*, 593–594. [CrossRef] [PubMed]
8. Hildebrandt, J.; O'Keeffe, S.; Bezama, A.; Thrän, D. Revealing the environmental advantages of industrial symbiosis in wood-based bioeconomy networks—An assessment from a life cycle perspective. *J. Ind. Ecol.* **2018**. [CrossRef]
9. Kruse, A.; Funke, A.; Titrici, M.M. Hydrothermal conversion of biomass to fuels and energetic materials. *Curr. Opin. Chem. Biol.* **2013**, *17*, 515–521. [CrossRef] [PubMed]
10. Libra, J.A.; Ro, K.S.; Kammann, C.; Funke, A.; Berge, N.; Neubauer, Y.; Titrici, M.M.; Fühner, C.; Bens, O.; Kern, J.; Emmerich, K.H. Hydrothermal carbonization of biomass residuals: A comparative review of the chemistry, processes and applications of wet and dry pyrolysis. *Biofuels* **2011**, *2*, 89–124. [CrossRef]
11. Greve, T.; Neudeck, D.; Rebling, T.; Röhrdanz, M. Prospects for the sustainable utilization of organic waste by Hydrothermal Carbonization. *Müll Abfall* **2014**, *2*, 86–93. Available online: https://www.muellundabfall.de/MA.02.2014.086 (accessed on 8 October 2018). (In German)
12. Fiori, L.; Lucian, M. Hydrothermal Carbonization of Waste Biomass: Process Design, Modeling, Energy Efficiency and Cost Analysis. *Energies* **2017**, *10*, 211. [CrossRef]
13. Lu, X.; Jordan, B.; Berge, N.D. Thermal conversion of municipal solid waste via hydrothermal carbonization: Comparison of carbonization products to products from current waste management techniques. *Waste Manag.* **2012**, *32*, 1353–1365. [CrossRef]
14. Zhang, Y. Hydrothermal Liquefaction to Convert Biomass into Crude Oil. In *Biofuels from Agricultural Wastes and Byproducts*; Hans, P., Blaschek, H.P., Ezeji, T.C., Scheffran, J., Eds.; Wiley-Blackwell: Hoboken, NJ, USA, 2010; pp. 201–232. ISBN 978-0-813-80252-7.
15. Kruse, A. Hydrothermal biomass gasification. *J. Supercrit. Fluids* **2009**, *47*, 391–399. [CrossRef]

16. Pala, M.; Kantarli, I.C.; Buyukisik, H.B.; Yanik, J. Hydrothermal carbonization and torrefaction of grape pomace: A comparative evaluation. *Bioresour. Technol.* **2014**, *161*, 255–262. [CrossRef]
17. Berge, N.D.; Flora, J.R.V.; Drive, B.; Carolina, N. *Energy Source Creation from Diverted Food Wastes via Hydrothermal Carbonization*; Technical Report; Environmental Research and Education Foundation: Raleigh, NC, USA, 2015.
18. Hognon, C.; Delrue, F.; Texier, J.; Grateau, M.; Thiery, S.; Miller, H.; Roubaud, A. Comparison of pyrolysis and hydrothermal liquefaction of Chlamydomonas reinhardtii. Growth studies on the recovered hydrothermal aqueous phase. *Biomass Bioenergy* **2015**, *73*, 23–31. [CrossRef]
19. De Mena Pardo, B.; Doyle, L.; Renz, M.; Salimbeni, A. (Eds.) *Industrial Scale Hydrothermal Carbonization: New Applications for Wet Biomass Waste*; ttz Bremerhaven: Bremerhaven, Germany, 2016; ISBN 978-3-00-052950-4.
20. Kruse, A.; Dahmen, N. Hydrothermal biomass conversion—Quo vadis? *J. Supercrit. Fluids* **2018**, *134*, 114–123. [CrossRef]
21. Elliott, D.C.; Hart, T.R.; Neuenschwander, G.G.; Rotness, L.J.; Olarte, M.V.; Zacher, A.V. Chemical Processing in High-Pressure Aqueous Environments. 9. Process Development for Catalytic Gasification of Algae Feedstocks. *Ind. Eng. Chem. Res.* **2012**, *51*, 10768–11077. [CrossRef]
22. Elliot, D.C.; Neuenschwander, G.G.; Hart, T.R.; Rotness, L.J.; Zacher, A.H.; Santosa, T.M.; Valkenburg, C.; Jones, S.B.; Tjokro Rahardjo, S.A. *Catalytic Hydrothermal Gasification of Lignin-Rich Biorefinery Residues and Algae*; Final Report; Pacific Northwest National Laboratory: Richland, WA, USA, 2009.
23. Jones, S.; Zhu, Y.; Anderson, D.; Hallen, R.; Elliot, D.; Schmidt, A.; Albrecht, K.; Hart, T.; Butcher, M.; Drennan, C.; et al. *Process Design and Economics for the Conversion of Algal Biomass to Hydrocarbons: Whole Algae Hydrothermal Liquefaction and Upgrading*; Pacific Northwest National Laboratory: Richland, WA, USA, 2014.
24. TerraNova Energy GmbH. Projekt Jining: TerraNova Ultra Anlage für 40.000 Jahrestonnen Klärschlamm. Available online: http://terranova-energy.com/blog/project/projekt-compiegne/ (accessed on 29 November 2018).
25. The Ingelia Patented HTC Plant. Available online: https://ingelia.com/index.php/modelo-negocio/carbonizacion-de-biomasa/?lang=en (accessed on 10 October 2018).
26. REVATECH Produkte Blue Coal. Available online: http://revatec.de/bluecoal.htm (accessed on 8 October 2018).
27. Klemm, M. Hydrothermale Carbonisierung (HTC) als Möglichkeit zur Klärschlammnutzung und Phosphorrückgewinnung. Presentation, 2016. Available online: http://www.thermolyphos.de/wp-content/uploads/2016/11/1-8-Klemm_161004_Hydrothermal.pdf (accessed on 8 October 2018).
28. Siemon, D.; HTC-Verfahren für Grünabfälle der SunCoal Industries. Berlin, Germany. Presentation, 2013. Available online: https://www.kompetenz-wasser.de/wp-content/uploads/2017/05/htc-4-von_ploetz.pdf (accessed on 08 October 2018). (In German)
29. TerraNova Energy. Clean Energy beyond Coal: Energiebilanz TerraNova Energy Verfahren. Presentation, 2011. Available online: https://www.landwirtschaftskammer.de/duesse/znr/pdfs/2011/2011-06-30-htc-05.pdf (accessed on 8 October 2018).
30. Umweltbundesamt (UBA) (Ed.) *Chancen und Risiken des Einsatzes von Biokohle und Anderer "Veränderter" Biomasse als Bodenhilfsstoffe oder für die C-Sequestrierung in Böden*; Umweltbundesamt: Dessau-Roßlau, Germany, 2016; p. 13ff.
31. Child, M. Industrial-Scale Hydrothermal Carbonization of Waste Sludge Materials for Fuel Production. Master Thesis, Lappeenranta University of Technology, Lappeenranta, Finland, 4 June 2014.
32. AVA-CO2. Klärschlamm Nutzen—Energie Gewinnen: Hydrothermale Carbonisierung. Available online: https://compa.pure-bw.de/sites/default/files/co_pdf_downloads/ava-co2_broschuere_de.pdf (accessed on 8 October 2018).
33. TerraNova Energy. Biokohle aus Reststoffbiomasse und Klärschlamm durch Hydrothermale Karbonisierung (HTC). Presentation, 2011. Available online: http://www.fh-meschede.de/einrichtungen/energietag/2011/pdf/Hydrothermale_Karbonisierung.pdf (accessed on 8 October 2018).
34. Sammeck, J. Wirtschaftlichkeit und Einsatzstoffe für HTC-Anlagen-erste Erfahrungen. Presentation, 2014. Available online: https://docplayer.org/8461905-Wirtschaftlichkeit-und-einsatzstoffe-fuer-htc-anlagen-erste-erfahrungen-narotec.html (accessed on 8 October 2018).

35. Weidner, E.; Elsner, P. (Eds.) *Bioenergie: Technologiesteckbriefe zur Analyse "Flexibilitätskonzepte für die Stromversorgung 2050"*; Schriftenreihe Energiesysteme der Zukunft: Berlin, Germany; München, Germany, 2016.
36. Reißmann, D.; Thrän, D.; Bezama, A. Techno-economic and environmental suitability criteria of hydrothermal processes for treating biogenic residues: A SWOT analysis approach. *J. Clean. Prod.* **2018**, *200*, 293–304. [CrossRef]
37. Rowe, G.; Wright, G. The Delphi technique as a forecasting tool: Issues and analysis. *Int. J. Forecast.* **1999**, *15*, 353–375. [CrossRef]
38. Sackman, H.; Delphi Assessment: Expert Opinion, Forecasting and Group Process. A Report Prepared for United States Air Force Project RAND. Santa Monica, CA, USA. 1974. Available online: https://www.rand.org/content/dam/rand/pubs/reports/2006/R1283.pdf (accessed on 14 September 2018).
39. Azman, H.; Mohd, N.H.F.; Mohd, S.M.H. Application of Fuzzy Delphi Approach Determining Element in Technical Skills among Students towards the Electrical Engineering Industry Needs. *Pertanika J. Soc. Sci. Hum.* **2017**, *25*, 1–8.
40. Wu, K.-Y. Applying the Fuzzy Delphi Method to Analyze the Evaluation Indexes for Service Quality after Railway Re-Opening—Using the Old Mountain Line Railway as an Example. *Recent Res. Syst. Sci.* **2011**, *1*, 474–479.
41. Kosko, B. Fuzzy cognitive maps. *Int. J. Man-Mach. Stud.* **1986**, *24*, 65–75. [CrossRef]
42. Salmeron, J.L.; Vidal, R.; Mena, A. Ranking Fuzzy Cognitive Maps based scenarios with TOPSIS. *Expert Syst. Appl.* **2012**, *39*, 2443–2450. [CrossRef]
43. Felix, G.; Napoles, G.; Falcon, R.; Froelich, W.; Vanhoof, K.; Bello, R. A review on methods and software for fuzzy cognitive maps. *Artif. Intell. Rev.* **2017**. [CrossRef]
44. Umweltbundesamt Austria. Fuzzy Cognitive Mapping—Das Werkzeug. Available online: http://www.umweltbundesamt.at/umweltsituation/oekosystemareumweltkontrolle/mfrp_eisenwurzen/projekte_eisenw/soz_oek_forsch/fcm/fcm_hinter/ (accessed on 9 October 2018).
45. Landeta, J.; Barrutia, J.; Lertxundi, A. Hybrid Delphi: A methodology to facilitate contribution from experts in professional contexts. *Technol. Forecast. Soc. Chang.* **2011**, *78*, 1629–1641. [CrossRef]
46. Gray, S.A.; Gray, S.; Cox, L.J.; Henly-Shepard, S. Mental Modeler: A Fuzzy-Logic Cognitive Mapping Modeling Tool for Adaptive Environmental Management. In Proceedings of the IEEE 46th Hawaii International Conference on System Sciences, Wailea, Maui, HI, USA, 7–10 January 2013. [CrossRef]
47. Stevenson, V. Some initial methodological considerations in the development and design of Delphi Surveys. In *H-Delivery WP 3—Task 3.2: Characterisation of Prospective Technologies*; Low Carbon Research Institute: Cardiff, UK, 2010; Available online: https://orca.cf.ac.uk/9949/1/Initial%20methodological%20considerations%20in%20the%20development%20and%20design%20of%20Delphi%20surveys.pdf (accessed on 26 November 2018).
48. Hasson, F.; Keeney, S.; McKenna, H. Research guidelines for the Delphi survey technique. *J. Adv. Nurs.* **2000**, *32*, 1008–1015. [CrossRef]
49. Cheng, C.H.; Lin, Y. Evaluating the best main battle tank using fuzzy decision theory with linguistic criteria evaluation. *Eur. J. Oper. Res.* **2002**, *142*, 174–186. [CrossRef]
50. Gray, S.; Zanre, E.; Gray, S. Fuzzy Cognitive Maps as Representations of Mental Models and Group Beliefs. In *Fuzzy Cognitive maps for Applied Sciences and Engineering—From Fundamentals to Extensions and Learning Algorithms*; Papageorgiou, E.I., Ed.; Springer Publishing: Berlin, Germany, 2018.
51. Özesmi, U.; Özesmi, S.L. Ecological models based on people's knowledge: A multi-step fuzzy cognitive mapping approach. *Ecol. Model.* **2004**, *176*, 43–64. [CrossRef]
52. Eden, C.; Ackerman, F.; Cropper, S. The analysis of cause maps. *J. Manag. Stud.* **1992**, *29*, 309–323. [CrossRef]
53. Obiedat, M.; Samarasinghe, S.; Strickert, G. A New Method for Identifying the Central Nodes in Fuzzy Cognitive Maps using Consensus Centrality Measure. In Proceedings of the 19th International Congress on Modelling and Simulation, Perth, Australia, 12–16 December 2011; pp. 1084–1091.
54. Tavassoli, S. Analyzing Centrality Indices in Complex Networks: An Approach Using Fuzzy Aggregation Operators. Ph.D. Thesis, Kaiserslautern, Germany, 2018.
55. Dahlman, C.J. The Problem of Externality. *J. Law Econ.* **1979**, *22*, 141–162. [CrossRef]
56. Kambo, H.S.; Minaret, J.; Dutta, A. Process Water from the Hydrothermal Carbonization of Biomass: A Waste or a Valuable Product? *Waste Biomass Valoriz.* **2018**, *9*, 1181–1189. [CrossRef]

57. Stutzenstein, P.; Weiner, B.; Köhler, R.; Pfeifer, C.; Kopinke, F.D. Wet oxidation of process water from hydrothermal carbonization of biomass with nitrate as oxidant. *Chem. Eng. J.* **2018**, *339*, 1–6. [CrossRef]
58. Köchermann, J.; Görsch, K.; Wirth, B.; Mühlenberg, J.; Klemm, M. Hydrothermal carbonization: Temperature influence on hydrochar and aqueous phase composition during process water recirculation. *Environ. Chem. Eng.* **2018**, *6*, 5481–5487. [CrossRef]
59. Cordell, D.; White, S. Peak Phosphorus: Clarifying the Key Issues of a Vigorous Debate about Long-Term Phosphorus Security. *Sustainability* **2011**, *3*, 2027–2049. [CrossRef]
60. Hsu, Y.-L.; Lee, C.-H.; Kreng, V.B. The application of Fuzzy Delphi Method and Fuzzy AHP in lubricant regenerative technology selection. *Expert Syst. Appl.* **2010**, *37*, 419–425. [CrossRef]
61. Kokkinos, K.; Lakioti, E.; Papageorgiou, E.; Moustakas, K.; Karayannis, V. Fuzzy Cognitive Map-Based Modeling of Social Acceptance to Overcome Uncertainties in Establishing Waste Biorefinery Facilities. *Front. Energy Res.* **2018**, *6*, 1–17. [CrossRef]
62. Helms, M.M.; Nixon, J. Exploring SWOT analysis—Where are we now? A review of academic research from the last decade. *J. Strategy Manag.* **2010**, *3*, 215–251. [CrossRef]
63. Rauch, P.; Wolfsmayr, U.J.; Borz, S.A.; Triplat, M.; Krajnc, N.; Kolck, M.; Oberwimmer, R.; Ketikidis, C.; Vasiljevic, A.; Stauder, M.; et al. SWOT analysis and strategy development for forest fuel supply chains in South East Europe. *For. Policy Econ.* **2015**, *61*, 87–94. [CrossRef]
64. Zadeh, L.A. Fuzzy sets. *Inf. Control* **1965**, *8*, 338–353. [CrossRef]
65. Hax, A.C.; Majluf, N.S. Competitive cost dynamics: The experience curve. *Interfaces* **1982**, *12*, 50–61. [CrossRef]

Article

Comparative Life Cycle Assessment of HTC Concepts Valorizing Sewage Sludge for Energetic and Agricultural Use

Kathleen Meisel [1,*], Andreas Clemens [1], Christoph Fühner [2], Marc Breulmann [2], Stefan Majer [1] and Daniela Thrän [1,3]

1 Deutsches Biomasseforschungszentrum Gemeinnützige GmbH (DBFZ), Bioenergy Systems Department, Torgauer Straße 116, 04347 Leipzig, Germany; andreas.clemens@tiptop-elbe.de (A.C.); stefan.majer@dbfz.de (S.M.); daniela.thraen@ufz.de (D.T.)
2 Helmholtz Centre for Environmental Research (UFZ), Environmental and Biotechnology Centre, Permoserstraße 15, 04318 Leipzig, Germany; chr.fuehner@gmail.com (C.F.); marc.breulmann@ufz.de (M.B.)
3 Helmholtz Centre for Environmental Research (UFZ), Department of Bioenergy, Permoserstraße 15, 04318 Leipzig, Germany
* Correspondence: kathleen.meisel@dbfz.de; Tel.: +49-341-2434-472

Received: 15 January 2019; Accepted: 20 February 2019; Published: 26 February 2019

Abstract: In many countries, sewage sludge is directly used for energy and agricultural purposes after dewatering or digestion and dewatering. In recent years, there has been a growing interest in additional upstream hydrothermal carbonization (HTC), which could lead to higher yields in the energetic and agricultural use. Twelve energetic and agricultural valorization concepts of sewage sludge are defined and assessed for Germany to investigate whether the integration of HTC will have a positive effect on the greenhouse gas (GHG) emissions. The study shows that the higher expenses within the HTC process cannot be compensated by additional energy production and agricultural yields. However, the optimization of the HTC process chain through integrated sewage sludge digestion and process water recirculation leads to significant reductions in GHG emissions of the HTC concepts. Subsequently, nearly the same results can be achieved when compared to the direct energetic use of sewage sludge; in the agricultural valorization, the optimized HTC concept would be even the best concept if the direct use of sewage sludge will no longer be permitted in Germany from 2029/2032. Nevertheless, the agricultural valorization concepts are not generally advantageous when compared to the energetic valorization concepts, as it is shown for two concepts.

Keywords: hydrothermal carbonization (HTC); life cycle assessment (LCA); sewage sludge; electricity and heat production; agricultural yield

---

1. Introduction

Sewage sludge is a by-product of wastewater treatment and it is produced in increasing quantities worldwide. 1.8 Mio. Mg dry matter of sewage sludge from municipal sewage treatment plants is produced in Germany, per year. Of this, 59.8 wt% is incinerated in coal-fired power plants and mono-combustion plants. 40.2 wt% is used in agriculture and landscaping [1]. In the case of energetic use, electricity and heat are generated, which predominantly substitute fossil energy in the German electricity and heating mix. Applying sewage sludge in agriculture can replace synthetically produced mineral fertilizers, which mostly rely on fossil-derived energy. Thus, both applications have the potential to contribute to climate protection.

For several years, there have been attempts to make sewage sludge treatment more efficient while using the Hydrothermal Carbonization (HTC) process. HTC is a thermochemical process that converts biomass into a solid product (hydrochar) at reaction conditions of 180 to 250 °C, 10 to 40 bar, and a pH of 3 to 7. One characteristic of HTC is that the conversion process takes place in a liquid, aqueous phase [2,3]. This means that wet, biogenic residues, such as biogenic municipal waste [4–7], fermentation residues [8,9], waste from agriculture and the food industry [10–15], as well as mechanically dehydrated sewage sludge with a water content of 75 wt% are particularly suitable as HTC feedstocks [16–18]. Like sewage sludge, hydrochars that are produced during HTC can be used for energy and agricultural purposes. For energetic use, the fact that hydrochar is easier to dewater when compared to the untreated sewage sludge is of particular importance. This positive effect has been extensively studied and it is described in [19–23]. For agricultural use, the hydrochar incorporation into soil is intended to improve plant growth and thus achieve higher crop yields. The soil-related use of hydrochar and its impact on plant growth have been described, among others, in [15,24–34].

In addition, the HTC of residue, waste, and biomass flows, such as green waste [6], olive mill waste [35], food wastes [36], biomass waste streams [37,38], and algae [39] has already been assessed several times by means of life cycle assessment (LCA). However, publications on HTC's LCA of sewage sludge are scarce. The HTC of bio-wastes, including sewage sludge, was investigated in [40]. Here, the focus was on energetic utilization in the form of fuel gas and diesel oil. In [41], different sewage sludge management methods were compared. The hydrothermal pyrolysis is one option investigated. The integration of HTC into an energetic utilization chain of sewage sludge is investigated in [42]. A comparative life cycle assessment, which includes the use of both sewage sludge and hydrochar, in energetic as well as agricultural applications, has not yet been conducted, to our knowledge.

In our study, while taking into account the processes of HTC with and without sewage sludge digestion as well as energetic and agricultural utilization, twelve different valorization concepts of sewage sludge are examined based on empirical and literature data. The aim is to identify the most promising concept from a greenhouse gas (GHG) perspective. In addition, it shall be determined (i) whether the integration of HTC with and without digestion in sewage sludge valorization concepts is more advantageous when compared to a direct utilization of sewage sludge and (ii) whether an energetic or an agricultural valorization is more favorable regarding their global warming potential.

2. Materials and Methods

2.1. Concept Definition

Twelve different valorizing concepts of sewage sludge are defined in this study; six with an agricultural use and six with an energetic use. Since there are sewage treatment plants with and without digestion of sewage sludge, both of the versions are considered in the concepts. To investigate whether the hydrothermal carbonization of sewage sludge offers advantages in comparison with direct agricultural or energetic use (as is mostly the case for Germany), the twelve concepts include both the direct use of sewage sludge (SS) after dewatering or digestion and dewatering and the use of hydrothermally carbonized sewage sludge (hydrochar). In the concepts of hydrothermal carbonization, the HTC process parameters are varied. In the first case (HTC 1), the HTC process runs at 170 °C and a residence time of 2 h, while in the second case (HTC 2), a temperature of 210 °C and a residence time of 10 h are chosen. The key characteristics of the concepts are also summarized in Table 1. Figure 1 shows an overview of the process concepts that were defined.

Table 1. Key characteristics of twelve valorization concepts investigated.

Process Chain	HTC Process Parameters	Agricultural Use	Energetic Use
SS without sludge digestion	-	A-SS	E-SS
SS with sludge digestion	-	A-D+SS	E-D+SS
HTC 1 without sludge digestion	170 °C, 2 h	A-HTC 1	E-HTC 1
HTC 2 without sludge digestion	210 °C, 10 h	A-HTC 2	E-HTC 2
HTC 1 with sludge digestion	170 °C, 2 h	A-D+HTC 1	E-D+HTC 1
HTC 1 with sludge digestion & recirculation of process water	170 °C, 2 h	A-D+HTC 1+RP	E-D+HTC 1+RP

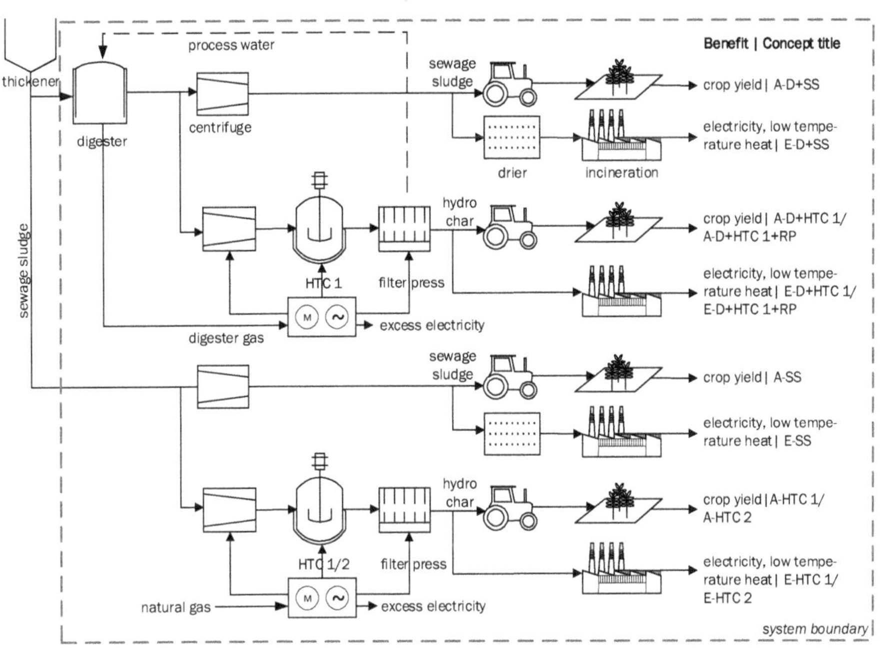

Figure 1. Overview of the valorization concepts investigated.

Since both options, with and without sewage sludge digestion, are considered, and all of the concepts shall be based on the same input material flow, the system boundary is defined after the thickener of the sewage treatment plant. The sewage sludge is assumed to leave the static thickener with a dry matter (DM) content of 5 wt%. The amount of sewage sludge is sized according to the capacity of an industrial HTC plant, set at 14,000 Mg a^{-1}, with a DM content of 25 wt%. This corresponds to a DM of 3500 Mg a^{-1} and thus to a quantity of sewage sludge from a size class 5 municipal sewage plant amounting to approximately 127,500 population equivalents (PEs). Thus, at DM content of 5 wt%, the amount of sewage sludge after the static thickener is 70,000 Mg a^{-1}.

The A-SS concept corresponds to the direct agricultural use of the non-anaerobically stabilized sewage sludge from the sewage treatment plant. The sewage sludge entering the system boundary is dewatered in a centrifuge to a DM content of 25 wt%. The centrate leaves the system boundary and it is returned to the sewage treatment plant, where it is mechanically, biologically, and chemically treated. The expenses for the wastewater treatment are included in all of the valorization concepts. The dewatered sewage sludge is transported to the field and then applied there. The effects of sewage sludge and hydrochar on plant growth were investigated by the cultivation of mustard (*Sinapis alba* L.), winter rye (*Secale cereale* L.), and corn (*Zea mays* L.) on a poor sandy site in Brandenburg (Germany). Starting the field trials with mustard in August 2014, already harvesting winter rye in May 2015 and cultivating corn in the summer of 2015, this rotation allowed for the cultivation of three crop species within only 12 months. The resulting yields leaves the system boundary as a benefit of the agricultural valorization concepts. Expenses that are required in the field, such as the seeds, pesticides, and diesel for agricultural machinery, are taken into account in the agricultural concepts. In the case of the concepts A-HTC 1 and A-HTC 2, a HTC plant is integrated into the process chain after the centrifuge. The sewage sludge, dewatered to 25 wt% of DM, is hydrothermally carbonized in the HTC plant and then dewatered in a filter press. The dewatered hydrochar has a DM content of 45.5 wt% (A-HTC 1) and 54.2 wt% (A-HTC 2), respectively. The filtrate leaves the system boundary with a DM content of 1.8 wt% (particle > 45 μm) and it is returned to the sewage treatment plant, where it is analogously cleaned to the centrate in the A-SS concept. A natural gas based combined heat and power (CHP) plant provides the HTC system with the required high-temperature heat and the auxiliary equipment with electricity. The excess electricity leaves the system boundary as a benefit. It is assumed that the excess heat cannot be further used. The hydrochar that is produced is transported to the field and applied to the soil. The two concepts, A-HTC 1 and A-HTC 2, only differ in terms of the parameters used in the HTC process (cf. Table 1).

In contrast to the concept A-SS, the concept A-D+SS integrates the anaerobic stabilization of the sewage sludge (digestion of sewage sludge) into the process chain. The incoming sewage sludge is anaerobically stabilized in the digester. Subsequently, the sewage sludge is dewatered in a centrifuge to a DM content of 25 wt%. Analogous to A-SS, the centrate is returned to the sewage treatment plant, where it is cleaned. The anaerobically stabilized and dewatered sewage sludge is transported to the field and then applied to the soil. Based on the concept A-D+SS, concepts A-D+HTC 1 and A-D+HTC 1+RP are defined. In both concepts, a HTC plant is integrated into the process chain after the centrifuge. The sewage sludge, dewatered to 25 wt% of DM, is hydrothermally carbonized in the HTC plant and then dewatered in a filter press. The hydrochar produced is again transported to the field and then applied to the soil. Both of the concepts differ only in the cleaning of the HTC process water. In A-D+HTC 1, the HTC process water leaves the system boundary and it is returned to the sewage treatment plant. In contrast, in A-D+HTC 1+PR the HTC process water is directed to the digester within the system boundary and anaerobically treated there. Consequently, more digester gas is produced when compared with the other concepts, including sludge digestion. The concept A-D+HTC 1+PR represents an optimized HTC concept. In the valorization concepts, including the sewage sludge digestion (A-D-SS, A-D+HTC 1, A-D-HTC 1+PR), the digester gas is fed to a CHP unit in order to produce heat and electricity. The CHP unit supplies the HTC plant with high-temperature heat and the digester with low-temperature heat. It is also assumed that the excess heat is not further used at the site of the sewage treatment plant and it leaves the system boundary as a loss. The electricity from the CHP unit supplies the electrical equipment. Excess electricity leaves the system boundary as a benefit.

The agricultural concepts are compared with corresponding energetic concepts. In this case, the sewage sludge or hydrochar are used to generate energy in a mono-combustion plant. According to [43], a lower heating value (LHV) of 4.5 MJ kg^{-1} is required for autothermal combustion in the mono-combustion plant. For this reason, to directly combust the sewage sludge, the sludge has to be dried thermally in a drier. This is done at the site of the mono-combustion plant, using the heat supplied from the mono-combustion plant. The condensate leaves the system boundary and then has to be cleaned. As already mentioned, the wastewater treatment is taken into account in the balancing. In contrast, the LHV of the hydrochar, which is mechanically dewatered in the filter press, is higher than 4.5 MJ kg^{-1}. This means that the HTC concepts do not need a thermal drier. In all energetic valorization concepts, the use of natural gas for the auxiliary firing, the use of an adsorbent (sodium hydrogen carbonate) for gas purification, and the disposal of the resulting ash are considered within the GHG assessment. A recycling of phosphorus and heavy metals from ashes is not considered. It is assumed that the excess heat that is produced from mono-combustion can be further used for other purposes or processes. Thus, the benefits of the energetic valorization concepts are electricity and low-temperature heat.

2.2. Calculation of GHG Emissions

To calculate and compare the GHG emissions of all the valorization concepts investigated, the life cycle assessment methodology according to the ISO guidelines 14040 and 14044 is applied [44,45]. Due to its bioenergy context, DIN ISO 13065 is also applied [46].

For this study, only the global warming potential (GWP 100), as expressed in CO_2 equivalents, is calculated as one of many possible environmental impact categories within the LCA framework. The characterization factors from the International Panel on Climate Change (IPCC) 2007 are used to convert all potential GHG emissions into the common unit of CO_2 equivalents [47]. Assuming that the sewage sludge consists entirely of biogenic material, the biogenic CO_2 emissions that were released during the combustion of the sewage sludge or hydrochar are considered to be climate-neutral in the energetic valorization concepts [48,49]. The same applies to the digester gas, which is incinerated in the CHP unit to provide the energy for the HTC process within the concepts with sludge digestion. In contrast, the direct carbon dioxide emissions from the combustion of natural gas in the CHP plant and in the mono-combustion plant (auxiliary firing) are included in the GHG balance.

In all of the valorization concepts, the main benefit is the disposal of sewage sludge. According to the German Institute for Standardization, the GHG emissions of concepts cannot be directly compared in order to identify the most favorable valorization concept, since they differ in their additional benefits (see Table 2) [44–46]. In the agricultural valorization concepts, agricultural yields are achieved, whereas in the energetic valorization concepts, electricity and heat are produced. This multifunctional problem can be addressed by means of different approaches [50–55]. In this study, each valorization concept is compared with its corresponding "substituted" reference system. To create these reference concepts, reference products, which fulfil the same benefits as the products in the corresponding valorization concept and may potentially be replaced by them, have to be identified. Within this substitution methodology, the choice of the "right" replaced production is decisive in fulfilling the adequate benefit [55,56].

Table 2. Main and additional benefits in the valorization and reference concepts; creation of equivalent benefits.

	Valorization Concepts	Reference Concepts
Agricultural Use		
Benefit	- 70,000 Mg a^{-1} of sewage sludge disposed of on the field - Mustard, rye, corn yield applying sludge and hydrochar - Excess electricity in HTC concepts	- 70,000 Mg a^{-1} of sewage sludge disposed of via co-combustion - Adequate mustard, rye, corn yield through mineral N fertilization - Electricity and low-temperature heat from co-combustion
Credits for	- Excess electricity from combined heat and power (CHP) unit at the HTC plant	- Electricity and low-temperature heat from co-combustion
Energetic Use		
Benefit	- 70,000 Mg a^{-1} of sewage sludge disposed of via mono-combustion - Electricity and low-temperature heat from mono-combustion - Excess electricity in HTC concepts	- 70,000 Mg a^{-1} of sewage sludge disposed of via co-combustion - Adequate generation of electricity and low-temperature heat via average power and low-temperature heat generation mix - Electricity and low-temperature heat from co-combustion
Credits for	- None; excess electricity from CHP plant is added as a benefit to electricity production from mono-combustion	- Electricity and low-temperature heat from co-combustion

In this study, the currently predominant conventional production is taken as reference production (see Table 2). The predominant conventional disposal of sewage sludge is its co-combustion in a power plant [1]. The expenses of co-combustion, such as dewatering of sewage sludge, ash disposal, and cleaning of the flue gas are therefore included in all of the reference concepts as adequate disposal of sewage sludge. During the co-combustion, electricity and heat are generated as an additional benefit in the reference concepts.

The benefit of the mustard, winter rye, and corn yields achieved in the agricultural valorization concepts is conventionally produced by mineral N-fertilization in the reference concepts. The electricity and low-temperature heat that are generated by the mono-combustion in the energy valorization concepts are assumed to replace the conventional German power and low-temperature generation mix in the corresponding reference concept.

To ensure that, as required in [44–46,57], the valorization and reference concepts have equivalent benefits, the products in the reference concepts have to fulfill the same benefits to the same extent and in the same quality as the products in the valorization concepts, but in a conventional manner. The substitution method is used to compensate for the differences in the benefits between the valorization and the reference concepts. Thus, all of the reference concepts get a GHG credit for the amount of electricity and heat generated by co-combustion (see Table 2). The same applies to the amount of excess electricity that is generated in the CHP process at the HTC plant in the valorization concepts. In the energetic valorization concepts, however, this amount of electricity is internally added as a benefit to the electricity that was generated by mono-combustion.

After creating equal benefits in the valorization and reference concepts, the GHG emissions are assessed for all concepts. In a next step, the GHG emissions from each valorization concept are subtracted from those from its corresponding reference concept to calculate the GHG saving that is associated with the potential substitution of each reference concept. Since all of the valorization concepts have the same annual input of 70,000 Mg a^{-1} of sewage sludge, the respective GHG savings, expressed in absolute figures in tones of CO_2 equivalents per year, can be compared and ranked. The valorization concept potentially causing the highest GHG saving as compared to its corresponding reference concept is the best concept from a global warming perspective.

2.3. Data and Assumptions

The characteristics of the sewage sludge and the produced hydrochars to be valorized in the concepts are shown in Table 3. The mass and energy balances for all concepts shown in Figure 1 are calculated within the defined system boundaries. These are listed in Tables 4–7. Efficiencies and further technical data for the processing units are taken from literature. The electrical efficiency of the natural gas CHP unit used in the HTC concepts is 38% [58], while that of the digester gas CHP unit is 30% [59]. Data from the HTC process comes both from the HTC reactor at a large-scale facility and from literature. The electrical efficiency and the fuel utilization rate of the mono-combustion are assumed to be 15% and 90%, respectively [60]. The co-combustion within the reference concepts is assumed to take place with an electrical efficiency of 42% and a fuel utilization rate of 44% [61]. The flows of the ash and flue gas from the mono- and co-combustion are taken from the energetic modelling of the combustion process.

Table 3. Characteristics of the sewage sludge and the hydrochars (wf = water free; waf = water and ash free).

Parameter	Unit	Sewage Suldge	Hydrochar HTC 1	Hydrochar HTC 2
Dry matter	wt%	25.0	45.5	54.2
LHV	MJ kg^{-1} $_{wf}$	11.64	11.63	13.58
Ash content	$wt\%_{wf}$	45.9	48.8	51.5
Elemental analysis				
C	$wt\%_{waf}$	51.8	55.3	63.7
H	$wt\%_{waf}$	7.1	7.0	8.0
S	$wt\%_{waf}$	1.8	3.8	4.4
N	$wt\%_{waf}$	7.5	7.8	6.6

Table 4. Overview of material balances of concepts for agricultural use.

Parameter	Unit	Without Digestion			With Digestion		
		A-SS	A-HTC 1	A-HTC 2	A-D+SS	A-D+ HTC 1	A-D+ HTC 1+ PR
Input							
Sewage sludge	Mg a^{-1}	70,000	70,000	70,000	70,000	70,000	70,000
Sulfuric acid	Mg a^{-1}	-	112	213	-	82	85
Natural gas CHP plant	Mg a^{-1}	-	461	494	-	-	-
Seed	Mg a^{-1}	114	60	59	114	60	66
Pesticides	Mg a^{-1}	6	3	3	6	3	4
Diesel	Mg a^{-1}	87	46	45	87	46	51
Output							
Mustard yield	Mg a^{-1}	1872	888	793	1869	887	981
Winter rye yield	Mg a^{-1}	1804	945	902	1801	944	1044
Corn yield	Mg a^{-1}	7380	3657	3847	7369	3652	4040
Centrate	Mg a^{-1}	56,045	56,045	56,045	59,009	59,009	65,549
HTC process water	Mg a^{-1}	-	9321	10,147	-	6786	-
Exhaust gas CHP plant CO_2 *	Mg a^{-1}	-	1249	1338	-	-	-

* Fossil CO_2 from the combustion of natural gas in the CHP plant.

Table 5. Overview of material balances of concepts for energetic use.

Parameter	Unit	Without Digestion			With Digestion		
		E-SS	E-HTC 1	E-HTC 2	E-D+SS	E-D+ HTC 1	E-D+ HTC 1+ PR
Input							
Sewage sludge	Mg a^{-1}	70,000	70,000	70,000	70,000	70,000	70,000
Sulfuric acid	Mg a^{-1}	-	112	213	-	82	85
Natural gas CHP plant	Mg a^{-1}	-	461	494	-	-	-
Natural gas auxiliary burner	Mg a^{-1}	2	1	1	1	1	1
Adsorbent gas cleaning	Mg a^{-1}	223	244	263	136	149	146
Output							
Centrate	Mg a^{-1}	56,045	56,045	56,045	59,009	59,009	65,549
HTC process water	Mg a^{-1}	-	9321	10,147	-	6786	-
Exhaust gas CHP plant CO_2 *	Mg a^{-1}	-	1249	1338	-	-	-
Condensate of drier	Mg a^{-1}	5169	-	-	4714	-	-
Exhaust gas from mono-inc. CO_2 **	Mg a^{-1}	122	130	140	75	80	79
Ash from mono-incineration	Mg a^{-1}	1284	831	876	1209	749	813

* Fossil CO_2 from the combustion of natural gas in the CHP plant; ** Fossil CO_2 from the combustion of natural gas in the auxiliary burner of the mono-incineration unit.

Table 6. Overview of energy balances of concepts for agricultural use.

Parameter	Unit	Without Digestion			With Digestion		
		A-SS	A-HTC 1	A-HTC 2	A-D+SS	A-D+ HTC 1	A-D+ HTC 1+ PR
Input							
Sewage sludge	MWh a^{-1}	16,235	16,235	16,235	16,235	16,235	16,235
Natural gas	MWh a^{-1}	-	6114	6552	-	-	-
Electricity	MWh a^{-1}	121	-	-	-	-	-
Output							
Electricity	MWh a^{-1}	-	1872	1531	1208	966	1154
Low-temp. heat	MWh a^{-1}	-	-	-	-	-	-

Table 7. Overview of energy balances of concepts for energetic use.

Parameter	Unit	Without Digestion			With Digestion		
		E-SS	E-HTC 1	E-HTC 2	E-D+SS	E-D+ HTC 1	E-D+ HTC 1+ PR
Input							
Sewage sludge	MWh a^{-1}	16,235	16,235	16,235	16,235	16,235	16,235
Natural gas	MWh a^{-1}	24	6127	6564	17	9	10
Electricity	MWh a^{-1}	121	-	-	-	-	-
Output							
Electricity	MWh a^{-1}	1289	2816	2721	1913	1514	1685
Low-temp. heat	MWh a^{-1}	3935	4718	5952	1235	2744	2654

Data regarding inputs to the agricultural valorization concepts, such as seeds, pesticides, diesel, and the yields of mustard, winter rye, and corn originate from field testings on the sandy marginal trial site in Brandenburg (Germany). The same field trials also provide the data on the agricultural inputs that are necessary to achieve equivalent mustard, winter rye, and corn yields via mineral fertilization in the agricultural reference concepts.

It is assumed that, in the agricultural valorization concepts, the treated sewage sludge and hydrochar are transported 30 km to the field, while in the energetic valorization concepts, both are transported 100 km to the mono-combustion plant.

For the GHG assessment, the mass and energy balances of the different valorization concepts are each transferred to a life-cycle model using the Umberto NXT Universal 7.1.13 software [62]. The emission data for the auxiliary materials and energy that are listed in the mass and energy balances come from the ecoinvent database v2.2 and v3.3 [63,64]. The direct and indirect nitrous oxide emissions resulting from the use of the treated sewage sludge and hydrochar in the field were determined in a laboratory using 40-day incubation experiments. The direct CO_2 emissions from the combustion of the natural gas in the CHP unit and the mono-combustion unit (auxiliary burner) are taken from our own balancing of the combustion process.

Creating the reference concepts, conventional co-combustion and conventional cultivation and energy generation systems are also modelled in Umberto. The emission factor that was used for the German electricity mix and low-temperature heat mix is sourced from the Gemis database v4.9 [65]. It is 0.61 kg CO_2-eq kWh^{-1} for power generation and 0.08 kg CO_2-eq. MJ^{-1} for the production of low-temperature heat [65]. These emission factors are also used to determine the credits for the additional generation of electricity and low-temperature heat.

3. Results

Figures 2 and 3 show clearly that only three agricultural valorization concepts (A-SS, A-D+SS, and A-D+HTC 1+PR) achieve GHG savings when compared to their corresponding conventional reference concepts. All other concepts that were investigated potentially emit more GHG emissions than their corresponding conventional reference concepts. In the concepts of direct valorization of sewage sludge (A-SS and A-D+SS), the applied sewage sludge contributes to achieving a relatively high yield of mustard, winter rye, and corn with low treatment expenses (dewatering or digestion and dewatering). If this yield is produced via mineral fertilizers as in the corresponding reference concepts, significantly more GHG are potentially emitted. This is mainly due to the higher GHG emissions from the production of synthetic nitrogen fertilizer when compared to the simple treatment of sewage sludge. In the agricultural valorization concepts A-HTC 1, A-HTC 2, and A-D+HTC 1, where the sewage sludge is converted to hydrochar via HTC, no GHG savings can be achieved as compared to the conventional reference systems. One reason is that the relatively high expenses of producing hydrochar in the HTC process cannot be compensated by higher yields when compared to the production system with conventional mineral fertilization.

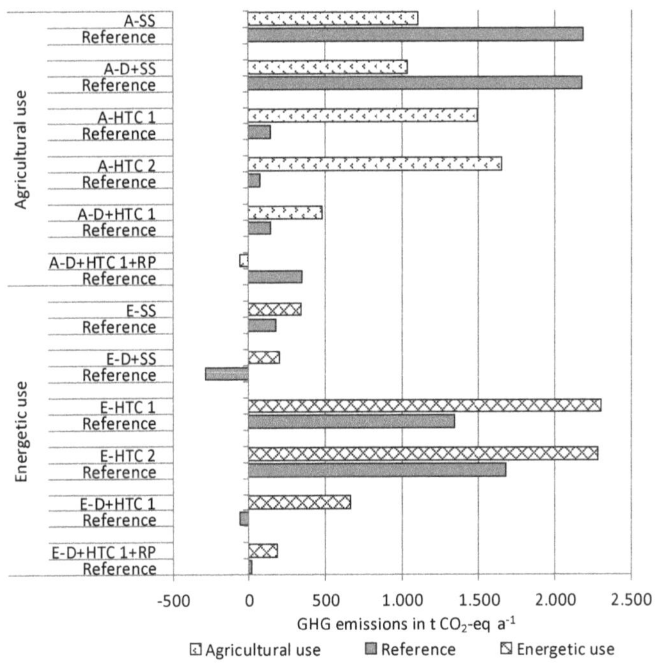

Figure 2. Greenhouse gas (GHG) emissions of valorization and reference concepts.

The other reason is that in all reference concepts of the agricultural valorization concepts, the conventional disposal of 70,000 Mg of sewage sludge (co-combustion in a power plant) is taken into account, in addition to the conventional production of the specific yield of mustard, rye, and maize. As well as requiring relatively low expenses, this co-combustion in the coal-fired power plant has an additional benefit as electricity and heat, which is credited to the reference concepts.

Despite these relatively high credits for the reference concepts, the optimized agricultural valorization concept, including HTC (A-D+HTC 1+PR), can reduce the GHG emissions when compared with the other agricultural valorization concept, including HTC (A-HTC 1 and A-HTC 2). When the process water is recirculated to the digester, additional digester gas is generated, which is used in the CHP unit instead of the fossil-derived natural gas and it produces a large amount of excess electricity, which is credited within the GHG balancing. As a result, the GHG emissions from the HTC process can be significantly reduced. It is even more than climate-neutral. Nevertheless, the GHG savings from the direct agricultural use of sewage sludge (A-SS, A-D+SS) are higher than from the optimized HTC concept when compared to their reference concepts.

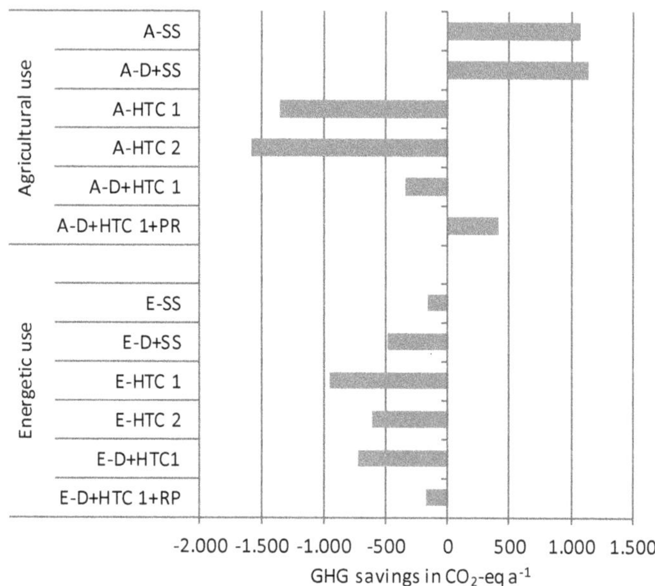

Figure 3. GHG savings comparing valorization and reference concepts.

By comparing the agricultural valorization concepts without digestion (A-SS, A-HTC1) and with digestion (A-D+SS, A-D+HTC1), it becomes clear that the concepts with digestion have potentially higher GHG saving or lower "additional GHG emissions" (see Figure 3). This is due to the generation of digester gas in the digester, which replaces the use of natural gas in the CHP unit.

Increasing the process parameters in the HTC process from 170 °C to 210 °C and from 2 h to 10 h residence time (A-HTC 2) leads to an increased demand for natural gas in the CHP unit but it does not adequately increase the yield of mustard, winter rye, and corn. The A-HTC 2 concept thus potentially causes more GHG emissions when compared to its reference concept than concept A-HTC 1 to its reference concept.

Figures 2 and 3 also show that all energetic valorization concepts cause more GHG emissions than their corresponding reference systems. The lowest "additional emissions" are caused by the concept of direct using the sewage sludge (E-SS), closely followed by the optimized concept with HTC, process water recirculation, and sewage sludge digestion (E-D+HTC1+PR). Both the concept E-SS and its corresponding reference system have similar expenses in the treatment of sewage sludge and mono- or co-combustion. Since the reference concept generates more electricity during co-combustion due to its higher electrical efficiency when compared with the mono-combustion in the valorization concept, the higher credit leads to lower GHG emissions in the reference concept. In the energetic valorization concepts, including HTC, the combustion of higher-grade hydrochar generates more power and heat as compared to the direct combustion of sewage sludge, but the higher expenses to produce hydrochar overcompensate for this positive effect. This trend is reduced in the concept E-D+HTC 1+PR. Here, again, recirculating the process water from the HTC process into the digester leads to a reduction in GHG emissions. However, this cannot compensate for the lower electricity generation when compared with the corresponding co-combustion in the reference concept. Nevertheless, the concept E-SS has the lowest additional emissions of all the energetic valorization concepts. In contrast to agricultural valorization concepts, sewage sludge digestion has no general beneficial effect on the energetic valorization. The concept E-D+SS has higher additional emissions as compared to its reference concept than the concept E-SS. Sewage sludge digestion, and thus the use of the digester gas instead of the natural gas in the CHP unit, reduces the GHG emissions in this valorization concept.

However, more energy, especially more heat, is generated from sewage sludge without their digestion in the mono-combustion unit. Analogous to the agricultural valorization concepts, the positive effect of sewage sludge digestion becomes clear when comparing the concepts E-HTC1 and E-D-HTC1. The reduced GHG emissions that are due to sewage sludge digestion outweigh the lower energy production in the mono-combustion.

In contrast to the agricultural valorization concepts, an increased carbonization temperature and HTC residence time (E-HTC 2) as compared to the E-HTC 1 concept leads to lower additional emissions, since the higher lower heating value of the hydrochar from the HTC 2 process leads to the generation of more electricity and heat.

Overall, the agricultural and energetic valorization concepts, including HTC, have no higher GHG savings when compared to its corresponding reference concepts than the concepts of direct use of sewage sludge. Within the agricultural valorization concepts, the optimized HTC concept A-D+HTC 1+PR potentially causes significantly less GHG saving than the concepts A-SS and A-D+SS, while, within the energetic valorization concept, the optimized HTC concept E-D+HTC 1+PR has nearly the same additional GHG emissions than E-SS as compared to their corresponding reference concepts (see Figure 3). Sewage sludge digestion, by contrast, has a positive effect on the GHG savings when compared to the reference concept, with just one exception (E-D+SS). Altogether, the agricultural valorization concept A-D+SS has the highest GHG saving, followed by A-SS and A-D+HTC 1+PR. Thus, the concept directly using the sewage sludge for agricultural purposes A-D+SS is the best concept from the global warming potential perspective. Only regarding the GHG emissions, the concept of A-D+HTC 1+PR causes the lowest GHG emissions and it is even more than climate-neural.

However, the agricultural valorization concepts are not generally advantageous when compared to the energetic valorization concepts, as in two cases (A-HTC 1 vs. E-HTC 1 and A-HTC 2 vs. E-HTC 2), the energetic valorization concepts would potentially emit less additional GHG emissions as compared to the reference concepts than the agricultural valorization concepts.

4. Discussion

The results show that only three of the agricultural valorization concepts can potentially achieve GHG savings when compared to their corresponding reference concepts, whereas all of the energetic valorization concepts emit more GHG emissions than their respective reference concepts. While the results of the energetic use of sewage sludge base on mass and energy balances that are derived from literature data and from a HTC reactor at a large-scale facility the mass and energy balances for the agricultural use are mainly based on data from a three-year test series (see Section 2.3) on sandy marginal revenue sites in Germany. Thus, the statements on GHG emissions and GHG savings of the agricultural valorization concepts, as well as the statements on the comparison between the agricultural and energetic valorization of sewage sludge, can only be transferred to a limited extent. There is a need for further research to generally determine the agricultural benefits of hydrochar, such as increases of crop yields or carbon sequestration in the soil over a long period.

The transferability of the results of the energetic valorization concepts is also limited, since they depend on the specific setting of the HTC process and reference systems defined. The results from [42] show that the advantageousness of HTC depends on the digester gas yields, the dewatering process during the HTC, the energy consumption of the HTC plants, and the combustion efficiencies.

In addition, the advantageousness of agricultural valorization reflected in this study is not reflected in the current German practice of sewage sludge utilization. Currently, more than half of sewage sludge in Germany is incinerated, and only about on-third is agriculturally used (see Introduction). One reason for the decline in the agricultural use of sewage sludge in recent years is that the thresholds for the concentration of cadmium, lead, nickel, and mercury that were permitted in the sewage sludge became stricter in the German Sewage Sludge and Fertiliser Ordinances [66,67]. Furthermore, the German Fertilizer Application Ordinance limits the application of organic fertilizers, including sewage sludge, to the fields [68]. Thus, the operator of sewage treatment plants primarily focus on energy valorization, which is legally less problematic.

When considering the changing legal regulations, the results of the study can be interpreted in another way. The German Directive on Reorganization of Sludge Valorization (2017) from 27 September 2017 prescribes the phosphorus recycling for sewage treatment plants (>100,000 PEs) from 2029 and for sewage treatment plants (>50,000 PEs) from 2032, which would no longer be possible via the direct co-combustion of sewage sludge in power plants, as the ashes would be diluted during co-combustion [69]. In addition, the direct agricultural use of sewage sludge is prohibited for these sewage treatment plants. However, sewage treatment plants of <50,000 PEs can still apply sewage sludge on soil. Thus, when considering the valorization of sewage sludge from sewage treatment plants >100,000 PEs, the first and second best concept for climate change mitigation cannot be applied from 2029 on. Assuming that the agricultural application of hydrochar will be allowed, the agricultural valorization concept A-D+HTC 1+PR is the best concept from a global warming potential perspective. If the application of hydrochar is also prohibited, the energetic valorization concepts E-SS and E-D+HTC 1+PR are the best sludge valorization options. However, it has to be mentioned that, leaving out co-combustion, will change the reference concepts and thus the GHG savings when compared to the reference.

In addition to the aspects that were investigated in this study, a number of other factors may also play a role in the decision on the integration of HTC into the concepts of valorization of sewage sludge. Some possible co-benefits of HTC are worth mentioning here, such as the degradation of organic pollutants, the removal of pathogens and helminths in the sewage sludge, and carbon sequestration in the soil, as well as the production costs that are associated with the valorization of sewage sludge.

5. Conclusions

The study shows that no fundamental advantage over concepts of direct agricultural and energetic valorization of sewage sludge regarding their global warming potential under the German condition are offered by integrating HTC into sewage sludge valorization concepts. The higher expenses within the HTC process cannot be compensated by additional agricultural yields and energy production. Only an optimization of the HTC process with integrated sewage sludge digestion and recirculation of the HTC process water enables comparable results. Altogether, the concept in which the sewage sludge is applied on the field after a relatively simple treatment of dewatering and digestion is the one with the highest GHG saving, and thus the most favorable concept regarding its impact on global warming. When the new requirements for valorizing sewage sludge come into effect from 2029 and 2032, neither direct agricultural use nor the direct co-combustion of sewage sludge from larger sewage treatment plants will be possible any longer in Germany. Thus, the best way to valorize sewage sludge would be to integrate a digestion of sewage sludge, as well as an optimized HTC process into the agricultural valorization concepts. Nevertheless, the agricultural valorization concepts are not generally advantageous when compared to the energetic valorization concepts, as it is shown for two concepts.

Author Contributions: Conceptualization, K.M., A.C. and C.F.; Data curation, A.C., C.F. and M.B.; Formal analysis, K.M. and A.C.; Methodology, K.M. and S.M.; Project administration, C.F.; Supervision, D.T.; Visualization, K.M.; Writing—original draft, K.M. and A.C.; Writing—review & editing, K.M., A.C., C.F., S.M. and D.T.

Funding: This research was funded by the Federal Ministry of Food and Agriculture (BMEL: 2815600211).

Conflicts of Interest: The authors declare no conflict of interest. The funder had no role in the design of the study; in the collection, analyses or interpretation of data; in the writing of the manuscript, or in the decision to publish the results.

References

1. Federal Statistical Office. Environment, Waste Water Treatment—Sewage Sludge. Final Report 2013/2014. Wiesbaden, 2017. Available online: https://www.destatis.de/DE/Publikationen/Thematisch/UmweltstatistischeErhebungen/Wasserwirtschaft/Klaerschlamm5322101139004.pdf?__blob=publicationFile (accessed on 16 February 2018).
2. Funke, A.; Ziegler, F. Hydrothermal carbonization of biomass: A summary and discussion of chemical mechanisms for process engineering. *Biofpr* **2010**, *4*, 160–177. [CrossRef]
3. Libra, J.A.; Ro, K.S.; Kammann, C.; Funke, A.; Berge, N.D.; Neubauer, Y.; Titirici, M.-M.; Fühner, C.; Bens, O.; Kern, J.; et al. Hydrothermal carbonization of biomass residuals: A comparative review of the chemistry, processes and applications of wet and dry pyrolysis. *Biofuels* **2011**, *2*, 71–106. [CrossRef]
4. Berge, N.D.; Ro, K.S.; Mao, J.; Flora, J.R.V.; Chappell, M.A.; Bae, S. Hydrothermal Carbonization of Municipal Waste Streams. *Environ. Sci. Technol.* **2011**, *45*, 5696–5703. [CrossRef] [PubMed]
5. Yao, Z.; Ma, X.; Lin, Y. Effects of hydrothermal treatment temperature and residence time on characteristics and combustion behaviors of green waste. *Appl. Therm. Eng.* **2016**, *104*, 678–686. [CrossRef]
6. Zeymer, M.; Meisel, K.; Clemens, A.; Klemm, M. Technical, Economic, and Environmental Assessment of the Hydrothermal Carbonization of Green Waste. *Chem. Eng. Technol.* **2017**, *40*, 260–269. [CrossRef]
7. Mumme, J.; Eckervogt, L.; Pielert, J.; Diakité, M.; Rupp, F.; Kern, J. Hydrothermal carbonization of anaerobically digested maize silage. *Bioresour. Technol.* **2011**, *102*, 9255–9260. [CrossRef] [PubMed]
8. Reißmann, D.; Thrän, D.; Bezama, A. Hydrothermal processes as treatment paths for biogenic residues in Germany: A review of the technology, sustainability and legal aspects. *J. Clean. Prod.* **2018**, *172*, 239–252. [CrossRef]
9. Zhao, X.; Becker, G.C.; Faweya, N.; Rodriguez Correa, C.; Yang, S.; Xie, X.; Kruse, A. Fertilizer and activated carbon production by hydrothermal carbonization of digestate. *Biomass Convers. Bioref.* **2018**, *8*, 423–436. [CrossRef]
10. Stemann, J.; Erlach, B.; Ziegler, F. Hydrothermal Carbonization of Empty Palm Oil Fruit Bunches: Laboratory Trials, Plant Simulation, Carbon Avoidance, and Economic Feasibility. *Waste Biomass Valoriz.* **2012**, *4*, 441–454. [CrossRef]
11. Basso, D.; Patuzzi, F.; Castello, D.; Baratieri, M.; Rada, E.C.; Weiss-Hortala, E.; Fiori, L. Agro-industrial waste to solid biofuel through hydrothermal carbonization. *Waste Manag.* **2016**, *47*, 114–121. [CrossRef] [PubMed]
12. Sabio, E.; Álvarez-Murillo, A.; Román, S.; Ledesma, B. Conversion of tomato-peel waste into solid fuel by hydrothermal carbonization: Influence of the processing variables. *Waste Manag.* **2016**, *47*, 122–132. [CrossRef] [PubMed]
13. Volpe, M.; Wüst, D.; Merzari, F.; Lucian, M.; Andreottola, G.; Kruse, A.; Fiori, L. One stage olive mill waste streams valorization via hydrothermal carbonization. *Waste Mang.* **2018**, *80*, 224–234. [CrossRef] [PubMed]
14. Mau, V.; Gross, A. Energy conversion and gas emissions from production and combustion of poultry-litter-derived hydrochar and biochar. *Appl. Energy* **2018**, *213*, 510–519. [CrossRef]
15. Zhang, S.; Zhu, X.; Zhou, S.; Shang, H.; Luo, J.; Tsang, D.C.W. Chapter 15—Hydrothermal Carbonization for Hydrochar Production and Its Application. In *Biochar from Biomass and Waste, Fundamentals and Applications*; Ok, Y.S., Tsang, D.C.W., Bolan, N., Novak, J.M., Eds.; Elsevier: Amsterdam, The Netherlands, 2019; pp. 275–294.
16. Robbiani, Z. Hydrothermal Carbonization of Biowaste/Fecal Sludge: Conception and Construction of a HTC Prototype Research Unit for Developing Countries. Master's Thesis, Swiss Federal Institute of Technology in Zurich (ETHZ), Zürich, Switzerland, 2013.
17. Danso-Boateng, E.; Holdich, R.G.; Shama, G.; Wheatley, A.D.; Sohail, M.; Martin, S.J. Kinetics of faecal biomass hydrothermal carbonisation for hydrochar production. *Appl. Energy* **2013**, *111*, 351–357. [CrossRef]
18. Parshetti, G.K.; Liu, Z.; Jain, A.; Srinivasan, M.P.; Balasubramanian, R. Hydrothermal carbonization of sewage sludge for energy production with coal. *Fuel* **2013**, *111*, 201–210. [CrossRef]
19. Namioka, T.; Morohashi, Y.; Yamane, R.; Yoshikawa, K. Hydrothermal Treatment of Dewatered Sewage Sludge Cake for Solid Fuel Production. *J. Environ. Eng.* **2009**, *4*, 68–77. [CrossRef]
20. Escala, M.; Zumbühl, T.; Koller, C.; Junge, R.; Krebs, R. Hydrothermal Carbonization as an Energy-Efficient Alternative to Established Drying Technologies for Sewage Sludge: A Feasibility Study on a Laboratory Scale. *Energy Fuels* **2013**, *27*, 454–460. [CrossRef]

21. Zhao, P.; Shen, Y.; Ge, S.; Yoshikawa, K. Energy recycling from sewage sludge by producing solid biofuel with hydrothermal carbonization. *Energy Convers. Manag.* **2014**, *78*, 815–821. [CrossRef]
22. Wang, L.; Li, A.; Chang, Y. Relationship between enhanced dewaterability and structural properties of hydrothermal sludge after hydrothermal treatment of excess sludge. *Water Res.* **2017**, *112*, 72–82. [CrossRef] [PubMed]
23. Kruse, A.; Dahmen, N. Hydrothermal biomass conversion: Quo vadis? *J. Supercrit. Fluids* **2018**, *134*, 114–123. [CrossRef]
24. Rillig, M.C.; Wagner, M.; Salem, M.; Antunes, P.M.; George, C.; Ramke, H.-G.; Titirici, M.-M.; Antonietti, M. Material derived from hydrothermal carbonization: Effects on plant growth and arbuscular mycorrhiza. *Appl. Soil Ecol.* **2010**, *45*, 238–242. [CrossRef]
25. Qayyum, M.F.; Steffens, D.; Reisenauer, H.P.; Schubert, S. Kinetics of carbon mineralization of biochars compared with wheat straw in three soils. *J. Environ. Qual.* **2012**, *41*, 1210–1220. [CrossRef] [PubMed]
26. Kammann, C.; Ratering, S.; Eckhard, C.; Müller, C. Biochar and hydrochar effects on greenhouse gas (carbon dioxide, nitrous oxide, and methane) fluxes from soils. *J. Environ. Qual.* **2012**, *41*, 1052–1066. [CrossRef] [PubMed]
27. Busch, D.; Kammann, C.; Grünhage, L.; Müller, C. Simple biotoxicity tests for evaluation of carbonaceous soil additives: Establishment and reproducibility of four test procedures. *J. Environ. Qual.* **2012**, *41*, 1023–1032. [CrossRef] [PubMed]
28. Gajić, A.; Koch, H.-J. Sugar beet (L.) growth reduction caused by hydrochar is related to nitrogen supply. *J. Environ. Qual.* **2012**, *41*, 1067–1075. [CrossRef] [PubMed]
29. George, C.; Wagner, M.; Kücke, M.; Rillig, M.C. Divergent consequences of hydrochar in the plant–soil system: Arbuscular mycorrhiza, nodulation, plant growth and soil aggregation effects. *Appl. Soil Ecol.* **2012**, *59*, 68–72. [CrossRef]
30. Malghani, S.; Gleixner, G.; Trumbore, S.E. Chars produced by slow pyrolysis and hydrothermal carbonization vary in carbon sequestration potential and greenhouse gases emissions. *Soil Biol. Biochem.* **2013**, *62*, 137–146. [CrossRef]
31. Bargmann, I.; Rillig, M.C.; Buss, W.; Kruse, A.; Kuecke, M. Hydrochar and biochar effects on germination of spring barley. *J. Agron. Crop Sci.* **2013**, *199*, 360–373. [CrossRef]
32. Puccini, M.; Ceccarini, L.; Antichi, D.; Seggiani, M.; Tavarini, S.; Hernandez Latorre, M.; Vitolo, S. Hydrothermal Carbonization of Municipal Woody and Herbaceous Prunings: Hydrochar Valorisation as Soil Amendment and Growth Medium for Horticulture. *Sustainability* **2018**, *10*, 846. [CrossRef]
33. Melo, T.M.; Bottlinger, M.; Schulz, E.; Leandro, W.M.; de Aguiar Filho, A.M.; Wang, H.; Ok, Y.S.; Rinklebe, J. Plant and soil responses to hydrothermally converted sewage sludge (sewchar). *Chemospehre* **2018**, *206*, 338–348. [CrossRef] [PubMed]
34. Breulmann, M.; van Afferden, M.; Müller, R.A.; Schulz, E.; Fühner, C. Process conditions of pyrolysis and hydrothermal carbonization affect the potential of sewage sludge for soil carbon sequestration and amelioration. *J. Anal. Appl. Pyrolysis* **2017**, *124*, 256–265. [CrossRef]
35. Benavante, V.; Fullana, A.; Berge, N.D. Life cycle analysis of hydrothermal carbonization of olive mill waste: Comparison with current management approaches. *J. Clean. Prod.* **2017**, *142*, 2637–2648. [CrossRef]
36. Berge, N.D.; Li, L.; Flora, J.R.V.; Ro, K.S. Assessing the environmental impact of energy production from hydrochar generated via hydrothermal carbonization of food wastes. *Waste Manag.* **2015**, *43*, 203–217. [CrossRef] [PubMed]
37. Owsianiak, M.; Ryberg, M.W.; Renz, M.; Hitzl, M.; Hauschild, M.Z. Environmental Performance of Hydrothermal Carbonization of Four Wet Biomass Waste Streams at Industry-Relevant Scales. *Acs Sustain. Chem. Eng.* **2016**, *4*, 6783–6791. [CrossRef]
38. Owsianiak, M.; Brooks, J.; Renz, M.; Laurent, A. Evaluating climate change mitigation potential of hydrochars: Compounding insights from three different Indicators. *Gcb Bioenergy* **2018**, *10*, 230–245. [CrossRef]
39. Patel, B.; Guo, M.; Izadpanah, A.; Shah, N.; Hellgardt, K. A review on hydrothermal pre-treatment technologies and environmental profiles of algal biomass processing. *Bioresour. Technol.* **2016**, *199*, 288–299. [CrossRef] [PubMed]
40. Svanström, M.; Patrick, T.N.; Fröling, M.; Peterson, A.A.; Tester, J.W. Choosing between green innovative technologies—Hydrothermal processing of biowastes. *J. Adv. Oxid. Technol.* **2007**, *10*, 177–185.

41. Lishan, X.; Tao, L.; Yin, W.; Zhilong, Y.; Jiangfu, L. Comparative life cycle assessment of sludge management: A case study of Xiamen, China. *J. Clean. Prod.* **2018**, *192*, 354–363. [CrossRef]
42. Remy, C.; Stüber, J. Weiterentwicklung des Klima- und Ressourceneffizienzpotentials Durch HTC-Behandlung Ausgewählter Berliner Klärschlämme—HTC-Berlin; Berlin, Germany, 2015. Available online: https://www.berlin.de/senuvk/umwelt/abfall/klaerschlamm/download/HTC-BERLIN_Abschlussbericht.pdf (accessed on 25 February 2018).
43. Schmitt, T.G.; Steinbrück, C.; Welker, A.; Dierschke, M. *Perspektiven einer zukunftsfähigen Klärschlammentsorgung in Rheinland-Pfalz*; Technische Universität Kaiserslautern: Kaiserslautern, Germany, 2007.
44. German Institute of Standardization. *DIN EN ISO 14040 Environmental Management—Life Cycle Assessment—Principles and Framework (ISO 14040:2006)*; Beuth Verlag: Berlin, Germany, 2006.
45. German Institute of Standardization. *DIN EN ISO 14044 Environmental Management—Life Cycle Assessment—Requirements and Guidelines (ISO 14044:2006)*; Beuth Verlag: Berlin, Germany, 2006.
46. German Institute of Standardization. *Sustainability Criteria for Bioenergy (ISO 13065:2015)*; Beuth Verlag: Berlin, Germany, 2015.
47. International Panel of Climate Change (IPCC). Climate Change 2007: The Physical Science Basis. Contribution of Working Group I to the Fourth Assessment Report of the Intergovernmental Panel on Climate Change, 2007. Available online: https://www.ipcc.ch/pdf/assessment-report/ar4/wg1/ar4_wg1_full_report.pdf (accessed on 15 February 2018).
48. Lombardi, L.; Nocita, C.; Bettazzi, E.; Fibbi, D.; Carnevale, E. Environmental comparison of alternative treatments for sewage sludge: An Italian case study. *Waste Manag.* **2017**, *69*, 365–376. [CrossRef] [PubMed]
49. Pawelzik, P.; Carus, M.; Hotchkiss, J.; Narayan, R.; Selke, S.; Wellisch, M.; Weiss, M.; Wicke, B.; Patel, M.K. Critical aspects in the life cycle assessment (LCA) of bio-based materials—Reviewing methodologies and deriving recommendations. *Resour. Conserv. Recycl.* **2013**, *73*, 211–228. [CrossRef]
50. Klöpffer, W.; Grahl, B. *Life Cycle Assessment (LCA): A Guide to Best Practice*; John Wiley & Sons: Weinheim, Germany, 2014; ISBN 3527655646.
51. Jung, J.; Von der Assen, N.; Bardow, A. Comparative LCA of multi-product processes with non-common products: A systematic approach applied to chlorine electrolysis technologies. *Int. J. Life Cycle Assess.* **2013**, *18*, 828–839. [CrossRef]
52. Ménard, J.-F.; Lesage, P.; Deschenes, L.; Samson, R. Comparative life cycle assessment of two landfill technologies for the treatment of municipal solid waste. *Int. J. Lca* **2004**, *9*, 371–378. [CrossRef]
53. Ekvall, T.; Finnveden, G. Allocation in ISO 14041—A critical review. *J. Clean. Prod.* **2001**, *9*, 197–208. [CrossRef]
54. Finnveden, G.; Hauschild, M.Z.; Ekvall, T.; Guinée, J.B.; Heijungs, R.; Hellweg, S.; Koehler, A.; Pennington, D.; Suh, S. Recent developments in Life Cycle Assessment. *J. Environ. Manag.* **2009**, *91*, 1–21. [CrossRef] [PubMed]
55. Heijungs, R.; Guinée, J.B. Allocation and 'what-if' scenarios in life cycle assessment of waste management systems. *Waste Manag.* **2007**, *27*, 997–1005. [CrossRef] [PubMed]
56. Weidema, B. Avoiding Co-Product Allocation in Life-Cycle Assessment. *J. Ind. Ecol.* **2001**, *4*, 11–33. [CrossRef]
57. Fleischer, G.; Schmidt, W.-P. Functional unit for systems using natural raw material. *Int. J. Lca* **1996**, *1*, 23–27. [CrossRef]
58. Arbeitsgemeinschaft für Sparsamen und Umweltfreundlichen Energieverbrauch e.V. (ASUE). BHKW-Kenndaten 2014/2015. Berlin, 2014. Available online: https://asue.de/sites/default/files/asue/themen/blockheizkraftwerke/2014/broschueren/05_10_14_bhkw_kenndaten_leseprobe.pdf (accessed on 3 February 2018).
59. Haberkern, B.; Maier, B.; Schneider, U. Steigerung der Energieeffizienz auf Kommunalen Kläranlagen. Dessau-Roßlau, 2008. Available online: https://www.umweltbundesamt.de/sites/default/files/medien/publikation/long/3347.pdf (accessed on 15 February 2018).
60. Obernberger, I.; Thek, G. Combustion and gasification of solid biomass for heat and power production in Europe – State-of-the-art and relevant future developments. In Proceedings of the 8th European Conference on Industrial Furnaces and Boilers, Dom Pedro Golf Resort, Vilamoura, 25–28 March 2008.
61. Füller, M. *Energie aus dem Nordosten*; Vattenfall Europe Mining & Generation: Cottbus, Germany, 2006.
62. Ifu Hamburg GmbH. *Umberto NXT Universal 7.1.13 Software*; Ifu Hamburg GmbH: Hamburg, Germany, 2017.

63. Ecoinvent Association. *Ecoinvent v2.2*; Swiss Federal Institute of Technology (ETH): Zurich, Switzerland, 2010.
64. Ecoinvent Association. *Ecoinvent v3.3*; Swiss Federal Institute of Technology (ETH): Zurich, Switzerland, 2016.
65. International Institute for Sustainability Analysis and Strategy (IINAS). *GEMIS-Global Emissions Modell for Integrated Systems v4.9*; International Institute for Sustainability Analysis and Strategy (IINAS): Darmstadt, Germany, 2014.
66. Verordnung über die Verwertung von Klärschlamm, Klärschlammgemisch und Klärschlammkompost (Klärschlammverordnung—AbfKlärV) from 27.09.2017. Available online: https://www.gesetze-im-internet.de/abfkl_rv_2017/AbfKl%C3%A4rV.pdf (accessed on 16 February 2018).
67. Verordnung über das Inverkehrbringen von Düngemitteln, Bodenhilfsstoffen, Kultursubstraten und Pflanzenhilfsmitteln (Düngemittelverordnung—DüMV) from 26.05.2017. Available online: https://www.gesetze-im-internet.de/d_mv_2012/DüMV.pdf (accessed on 16 February 2018).
68. Verordnung über die Anwendung von Düngemitteln, Bodenhilfsstoffen, Kultursubstraten und Pflanzenhilfsmitteln nach den Grundsätzen der Guten Fachlichen Praxis beim Düngen (Düngeverordnung—DüV) from 26.05.2017. Available online: https://www.gesetze-im-internet.de/d_v_2017/DüV.pdf (accessed on 3 February 2019).
69. Directive on Reorganization of Sludge Valorization from 27.09.2017. Available online: https://www.bgbl.de/xaver/bgbl/start.xav?startbk=Bundesanzeiger_BGBl&start=%2F%2F%2A%5B%40attr_id=%27bgbl117s3465.pdf%27%5D#__bgbl__%2F%2F*%5B%40attr_id%3D%27bgbl117s3465.pdf%27%5D__1549402355053 (accessed on 3 February 2019).

MDPI
St. Alban-Anlage 66
4052 Basel
Switzerland
Tel. +41 61 683 77 34
Fax +41 61 302 89 18
www.mdpi.com

Energies Editorial Office
E-mail: energies@mdpi.com
www.mdpi.com/journal/energies

www.ingramcontent.com/pod-product-compliance
Lightning Source LLC
LaVergne TN
LVHW070846170726
843515LV00005B/584
* 9 7 8 3 0 3 9 2 8 6 7 6 8 *